An Eventful Journey to
Unification of All the Fundamental Forces

Other Related Titles from World Scientific

Physics off the Beaten Track: A Selection of Papers by Robert Delbourgo
edited by Robert Delbourgo
ISBN: 978-981-3277-89-2

Classical Mechanics and Electrodynamics
by Jon Magne Leinaas
ISBN: 978-981-3279-36-0
ISBN: 978-981-3279-98-8 (pbk)

Group Theory for Physicists
Second Edition
by Zhong-Qi Ma
ISBN: 978-981-3277-38-0
ISBN: 978-981-3277-96-0 (pbk)

Loop Quantum Gravity for Everyone
by Rodolfo Gambini and Jorge Pullin
ISBN: 978-981-121-195-9

An Eventful Journey to
Unification of All the
Fundamental Forces

Robert Delbourgo

University of Tasmania, Australia

World Scientific

NEW JERSEY · LONDON · SINGAPORE · BEIJING · SHANGHAI · HONG KONG · TAIPEI · CHENNAI · TOKYO

Published by

World Scientific Publishing Co. Pte. Ltd.

5 Toh Tuck Link, Singapore 596224

USA office: 27 Warren Street, Suite 401-402, Hackensack, NJ 07601

UK office: 57 Shelton Street, Covent Garden, London WC2H 9HE

British Library Cataloguing-in-Publication Data
A catalogue record for this book is available from the British Library.

AN EVENTFUL JOURNEY TO UNIFICATION OF ALL THE FUNDAMENTAL FORCES

ISBN 978-981-121-014-3

For any available supplementary material, please visit
https://www.worldscientific.com/worldscibooks/10.1142/11546#t=suppl

Desk Editor: Ng Kah Fee

Typeset by Stallion Press
Email: enquiries@stallionpress.com

This book is dedicated to my son, Tino. Being an Honours graduate in Physics with many years experience at teaching children of all levels in secondary school, he was well equipped to advise what material is likely to be understandable to readers with some scientific background and a modicum of mathematical skills. Thus he served as an ideal sounding board/critic for knowing what stuff to include and what to leave out, what is easily comprehensible and what is not; in particular how much mathematics to include without making the book heavy going. When the material did not pass muster with him, it forced me back to the drawing board. I hope then that this final version of the monograph makes my journey to a novel picture of force unification accessible to fellow travellers.

Preface

There is little doubt that human beings, with their highly developed brains, have become the most inquisitive animals on this planet. Human curiosity did not blossom until the development of agriculture allowed for more leisure time to contemplate and investigate the environment and especially work out how the natural laws operate. At first this led to crude ideas about the function of various forces that inevitably got enmeshed with superstition, right until the scientific method was developed. In its turn this led to experimental probing of the forces and the overthrow of unbelievable tenets, so much so that our progress in understanding of nature's workings continues to gather pace.

The first step in the progression towards scientific enlightenment involved gathering information about the heavens, tides and seasons but with little comprehension about the underlying mechanisms. It was only after Galileo, Hooke and Newton established the laws of mechanics and gravitation that we could really appreciate our place in the universe, culminating in a proper understanding of the gravitational force. The second step took a lot longer; it involved uncovering the electromagnetic force and evolved sporadically from observation of electrostatic attraction/repulsion, most spectacularly from lightning discharges; on the parallel magnetic side it advanced from using lodestone as a practical navigational aid to discovering what happened when magnets and charges are set in motion. Once reliable batteries (among other voltage sources) were produced variable electromagnetic fields/forces and their associated currents lent themselves to proper scientific investigation; they culminated with their final mathematical encapsulation in Maxwell's famous equations.

Over the next fifty years the next steps emerged. They concerned the nature of matter, focussing on atomic structure and the nuclear forces which hold the nucleus together. That particular period saw the birth of exciting developments in quantum theory and relativity. The creation of accelerators or atom-smashers and the discovery of cosmic rays heralded this fresh field. An avalanche of new particles were unearthed one after another; this included mesons, strange particles, leptons and neutrinos. Mesons in the form of pions served to describe the strong short-range force between nucleons: pionic exchange was deemed by Yukawa to be responsible for nuclear binding, similarly to photonic exchange for atomic binding. It took a further fifty years to realise that nuclei were themselves made up of smaller constituents named quarks by Gell-Mann — an idea which went a long way towards explaining the proliferation of new particle combinations. Today we appreciate that these quarks are bound in a permanent embrace due to a 'chromodynamic' (QCD) force which acts on three colours carried by quarks; actually the gluonic exchange between quarks is much more analogous to electromagnetism (QED) where vector photons are exchanged between charged sources; this relegates scalar pions from a primary role into a secondary one.

That same period saw the emergence and recognition of the forces that produce radioactivity which cause element transmutation and which sparked a minor revolution in chemistry. They are named weak interactions based on their feebleness at ordinary energies and they are characterised by violation of mirror symmetry or 'parity' that surprised everyone. It took a while before weak forces became unified with electromagnetism into a so-called 'standard electroweak model' which robustly explains pretty well all the features and interactions of the particle multitude up to the present day.

During all that time, gravity, being really weak except at a macroscopic level, took a back seat. Even so, seminal discoveries kept on confirming predictions arising from Einstein's celebrated theory of general relativity (GR). GR overturned Newtonian concepts and gave them a geometric perspective. Today these gravitational and relativistic effects, while small, are of crucial importance; they have been exploited, for example in GPS positioning. The most exotic of these is the prediction of gravitational waves in a 'plastic' spacetime as was recently verified by high precision laser interferometry.[1] The marriage of relativity and quantum theory for gravity has unfortunately proved more problematic. Quantisation of gravity (previously carried out successfully in electroweak theory and QCD) can be achieved but is in serious difficulty through quantum effects which lead to untameable infinities.

This has not discouraged many researchers from following in Einstein's footsteps by seeking to geometrise electroweak theory, but so far without any resounding success. This book will sketch the history of these attempts after which I shall try to present my own ideas about producing a unified geometrical picture of these four fundamental forces. The scheme is based entirely on the complete description of events: 'where–when–what', the key being *what*. The very early parts of the book contain little mathematics and what there is should be accessible to general readers; these sections serve as a refresher course in modern physics. However the next few sections do require some knowledge of and familiarity with undergraduate mathematics/physics as they otherwise would become rather incomprehensible; I make no apology for that but will try at least to convey the general principles to those readers who baulk at mathematical formalism. In the last few sections some of the maths has been ramped up but at all times I have striven to make the meaning of seemingly complex equations/algebra more physically comprehensible. For those readers who have gone beyond freshman mathematics an Appendix has been included to summarise some of the formalism. I have also tried to humanise the enterprise in which we are all engaged by interspersing the text with small portraits of famous physicists that go back many decades. With physicists of the present or a somewhat earlier generation I have referred to their key papers.

---

[1]See the press release by LIGO: https://www.ligo.caltech.edu/page/press-release-gw170817.

Finally there is a basic Bibliography section and a Glossary for terms which recur.

This monograph therefore tries to provide a framework and picture for a 'theory of everything', however elusive the endpoint. Many approaches towards reaching that goal have been tried; while the ideas of those scientists are very imaginative and indeed quite brilliant, they remain stubbornly unconfirmed by experiment, if not ruled out. In retirement I have had the luxury of time and contemplation in trying a novel approach which is quite different from other attempts; those efforts have resulted in getting the main ideas published in refereed research journals from the World Scientific stable. More recently I was invited to write an account of that work in the AIP's 'Australian Physicist'. This monograph has grown out of that material and, quite obviously, previously published research. As stated above, the aim of the book is to exhibit those concepts at a level that any scientist with a physical science bachelor's degree can appreciate even if not in great mathematical depth. (For those readers with a good general mathematical knowledge, I invite them to go through the material and try to check the results as they are quite instructive.) The monograph contains some new results but, as with any scientific enterprise, they may be found wanting in the light of deeper research. If fatal cracks of this new approach for characterising events do emerge, the whole concept may prove to be another casualty of the 'nothing ventured, nothing gained' club, and in famous company!

Robert Delbourgo

Acknowledgments

After reading a semi-popular article on the subject of unified forces which appeared in the *Australian Physicist*, Professor C S Lim suggested that I should expand it into a monograph. With gentle persuasion he overcame my extreme hesitancy because I felt that the material was already covered in the research publications. It is my hope that this monograph helps to highlight the ideas contained in those papers for science graduates who have not pursued further studies in particle physics. Of course there exist gaps in my research, as in any other scientific topic; perhaps experts in quantum field theory will be sufficiently motivated to fill them in and venture to extend my approach. There are surely byways that I have not explored; of that I am convinced.

I need to acknowledge the help I have received in this endeavour from the following colleagues: Peter Jarvis, Roland Warner, Ruibin Zhang and Paul Stack. I am greatly indebted to them for their inputs. Brian Kenny has also supported me staunchly throughout that time. Last but not least I would like to credit Wikimedia Commons for the pictures of the physicists sprinkled in various places in the text.

Contents

Chapter 1

A Historical Look at Forces

In order to appreciate the advances in physical science that have taken place during the last fifty years it is worthwhile retracing the history of their development. Nowadays we are all agreed that four fundamental forces exist, so let us trace how such a conclusion was reached. I should add that there are some suggestions of a fifth force, connected with dark matter, but they are not supported at all convincingly by experiment, so I will not dip my feet into that pool.

1.1 Gravitational attraction

All animals on Earth become aware of gravity soon after their births. Gravity is essential to life and human well-being, as astronauts who have spent long periods in space have discovered to their detriment. Many animals exploit gravity to their advantage and humans unfortunately have exploited it for warfare, mainly to outslay their enemies, progressing by 'trial and error' throughout the centuries. A proper understanding of projectile motion had to await the correct formulation of mechanics by Galileo and Newton and a true understanding of frictional forces in various environments. It led to the entire field of ballistic science, aeronautics, space travel, etc.

Newton's three laws of motion were the stepping stones for completely describing the behaviour of celestial bodies. Hooke had appreciated that some sort of gravitational attraction between planets and the Sun (decreasing with increasing separation) was responsible for their Keplerian trajectories, but it took the genius of Newton, hand in hand with the formulation of calculus (or 'fluxions' as Newton

Fig. 1.1   Galileo Galilei.

Fig. 1.2   Johannes Kepler.

Fig. 1.3   Robert Hooke.

Fig. 1.4   Isaac Newton.

termed it), to establish the inverse square law of attraction, proving that it led to elliptical planetary orbits and was responsible for tides. That put astronomy on a firm mathematical footing and is the foundation for astronomical analyses today, despite the occasional intrusions of relativity.

One of the more puzzling aspects of Newton's universal gravitational law of attraction between two masses (M and m) is the appearance of the same mass parameter m in the equation

$$F_{\text{gravity}} = G_N M m / r^2$$

that enters Newton's second law

$$F_{\text{motion}} = d(mv)/dt.$$

To be sure, for a fixed mass, it leads to the same gravitational acceleration $g = dv/dt = G_N M / r^2$ of free fall, as Galileo is reputed to have tested at Pisa. But why should that be, regardless of the makeup of the body? It required Einstein's unique brilliant imagination to realise that this had nothing to do with the body itself but was a consequence of the geometry of the spacetime continuum in which the body resides. Thus was born the general theory of relativity (GR) and an entirely new view of the universe; the ramifications of Einstein's ideas about relativity have been tested successfully in many ways and continue to be checked today with ever more sensitive instrumentation. This new perspective also disposes of the Newtonian 'action at a distance' hypothesis, showing it to be invalid because the gravitational forces between bodies are in fact communicated (like electrical ones), at the speed of light, not infinitely fast. I shall return to the force of gravity later when discussing fields.

1.2 Electromagnetic influence

It is a curious circumstance that animals made use of magnetic forces well before humans. Birds, such as homing pigeons, steer their way on migration routes by cryptochromes in their eyes, while bats and even worms orient themselves by detecting the Earth's field. Monotremes make use of electricity to locate prey through electrical impulses; some marine animals such as sharks and rays use the same strategy underwater, while electric eels possess batteries strong enough to stun their prey.

Navigators had been using lodestones (magnetised magnetite) to direct their explorations over many centuries; crude compasses became a prominent navigational aid when the skies were overcast and it was no longer possible to use the Sun or stars. Gilbert was the first to properly understand that the Earth was acting like a huge magnet and this led him to conjecture that the core of our planet consisted mainly of iron. We must also point out that magnetism was one of the key indicators establishing plate tectonics in the history of geological discovery.

By contrast, apart from lightning discharges during violent storms, electrical force played a minor role in human history until about the eighteenth century. Voltages produced by some animals or in storms are hardly controllable and not conducive to scientific experimentation. Humans had been aware of the phenomenon of electrostatics, caused principally by friction, but mainly as a curiosity and not as a tool. Priestley and Coulomb were among the first to study electrostatics and enunciate the inverse square law of attraction/repulsion by making use of torsion balances; to their surprise and delight they found that the law resembles the Newtonian gravitational law. In parallel with that research, the connection between electricity and animal mobility was accidentally deduced by Galvani following observations of twitching froglegs held between two different metals; Volta correctly attributed this to electrochemical effects and produced the first Cu–Zn battery, which in turn led to discovery of several chemical elements.

Detailed investigation of current electricity could not proceed until the creation of reliable fixed voltage batteries, such as the Daniell and Leclanche cells, at which point research on this topic exploded. Electromagnetism entered a golden age of discovery pioneered by Oersted, Ampère, Biot-Savart, Faraday, Lenz and Ohm. The first discovery was that a moving charge, namely a current, can induce a magnetic force, and it was quickly followed by the converse

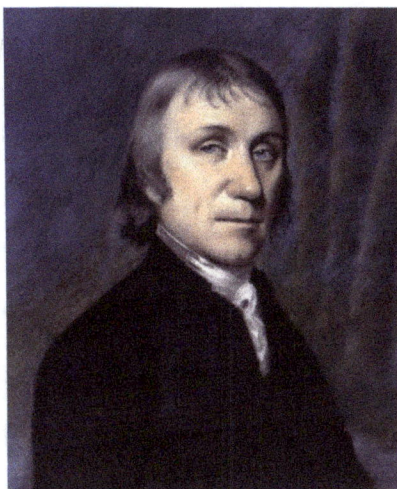

Fig. 1.5   Joseph Priestley.

Fig. 1.6   Charles-Augustin de Coulomb.

effect of electromagnetic induction whereby a moving magnet can lead to an electromotive force. Relying upon the interconnection between electricity and magnetism, the galvanometer became a crucial experimental tool for quantifying these phenomena. Eventually it led to the invention of electric motors and generators. A modern world is inconceivable without considering the benefits of electromagnetism; yet it is ironic to think that two hundred and fifty years ago we knew hardly anything about electromagnetic forces. We shall return to this topic presently with more stress on the concept of force 'fields'.

Aside from energy production/distribution issues (begun by Edison and Tesla), electromagnetic waves have played a very significant role in societal development. Beginning with radio wave transmission/reception at the start of the 20th century, speed of communication has caused a revolution in human progress. Micro-miniaturisation of electronic components such as the transistor has been an essential ingredient for this advance. Today we make use of the entire electromagnetic spectrum, radio to microwave through

Fig. 1.7   Luigi Galvani.

Fig. 1.8   Alessandro Volta.

Fig. 1.9   André Marie Ampère.

Fig. 1.10   Hans Christian Oersted.

Fig. 1.11   Michael Faraday.

Fig. 1.12   Heinrich Friedrich Emil Lenz.

infrared, visible to ultraviolet and on to X-ray and gamma rays. The impact this has had on electronic gadgetry cannot be overstated. Also the use of X-rays and gamma rays has led to medical diagnostic improvements and treatments. Furthermore the influence on astronomical discoveries has been huge, both qualitatively and quantitatively. It is true to say that in our modern society electromagnetism reigns supreme when compared to the other forces.

1.3   Nuclear bonding

The concept of a strong nuclear force had to wait until the start of the twentieth century when experiments revealing the structure of atoms became a dominant area of physics research. Rutherford and Marsden uncovered the planetary-like nature of elemental atoms through their investigations of bombarding α particle trajectories, the αs being supplied by radioactive sources. The revelation that the positively charged nucleus was some 100,000 times smaller than the atom itself (consisting of negatively charged circling electrons) caught

Fig. 1.13   Georg Simon Ohm.

Fig. 1.14   James Clerk Maxwell.

everyone by surprise; at first the mass of the nucleus relative to electrons proved to be a puzzle. It was only after Chadwick discovered the neutron through nuclear transmutations that the character of the central nucleus was elucidated. At the same time the meaning of atomic number was clarified and the curious spread of atomic mass became obvious with an understanding of isotopic composition. Mendeleev's periodic table came fully into its own.

Nuclear physics involves energies some million times larger than atomic physics. Investigating that subject needed the use of MeV accelerators — starting with Van de Graaff generators progressing to cyclotrons, betatrons, linacs, etc. Nuclei had to be smashed before they could yield the secrets of the nuclear force which binds protons and neutrons. Paul Matthews has likened this form of experimentation to the following analogy: the nucleus is like an invisible statue; we turn on a high pressure hose on it and try to discover the shape of the statue by the observed splashes. [In Figures 1.19 and 1.20 we contrast the scales and energies of the first particle accelerator, developed by Cockroft and Walton, with the modern Large Hadron Collider which operates in CERN.]

Fig. 1.15   Ernest Rutherford.

Fig. 1.16   Ernest Marsden.

Fig. 1.17   James Chadwick.

Fig. 1.18   Hideki Yukawa.

Fig. 1.19   A Cockroft–Walton particle accelerator operating at 10 MeV.

Most nuclei are stable because the strong nuclear attraction between protons and neutrons overcomes the electrical repulsion between the participating protons. This shows that another attractive mechanism is at work, very different from electromagnetism or gravity. It is very strong and only acts over short distances. Yukawa determined that unions between nucleons are governed by meson attraction in the same way that atoms are held together through electric attraction between electrons and positively charged nuclei.

Nuclear instability occurs when electric repulsion between protons in the nucleus is large enough to overcome the mesonic attraction. It is one of the principal causes of radioactivity. From studies of elemental (in)stabilities we now have a very good picture of nuclear species. Humans have exploited the relative longevity of some radioactive elements, such as uranium and thorium, to find energy supplies that were unimaginable a hundred years before. The energies produced are

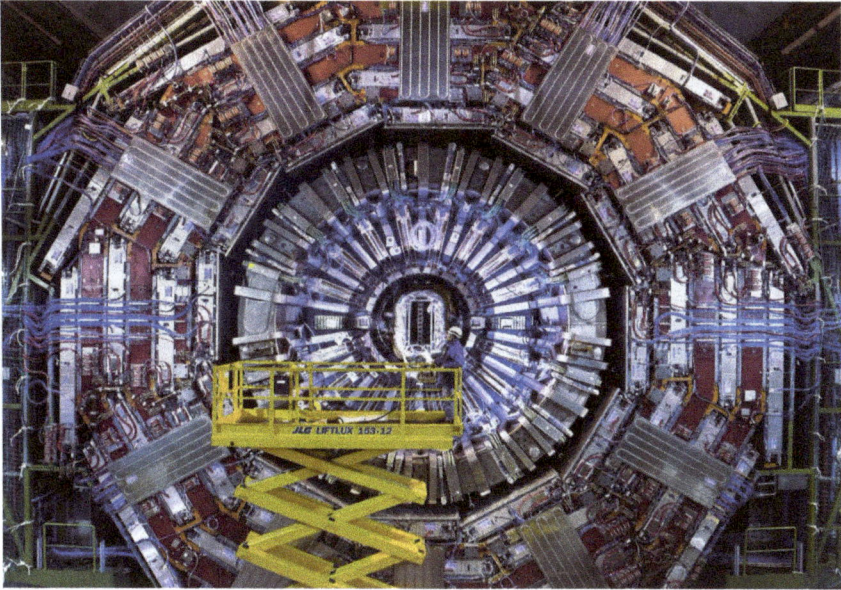

Fig. 1.20   CERN Large Hadron Collider operating at about 10 TeV. [By kind permission from CERN].

many orders of magnitude greater than those provided by chemical processes. The first use of these stupendous energies came through warfare and their taming took several years; secondarily it has led to nuclear reactors based on fissile materials and an entire energy production sector. The nuclear forces associated with fusion (as the Sun is powered) are more challenging to tame; this is because the plasma involved plus the heat produced need to be contained in some way without destroying the container. It has led to a vibrant area of research called magneto-hydrodynamics into which billions have been invested. When fusion is eventually mastered/controlled it is expected to lead to a cheap source of energy, namely heavy water held in the oceans.

1.4   Weak disintegration

Becquerel was the first person to discover radioactivity, almost by accident. He noticed that photographic plates were affected by

Fig. 1.21 Antoine Henri Becquerel.

Fig. 1.22 Pierre and Marie Curie.

pitchblende even though those plates were stored in a dark drawer. Radioactivity showed that certain elements are *not* forever, with the culprit being the decaying uranium/radium in pitchblende. It destroyed the old rule of chemistry that atoms are indestructible. Nuclear transmutation became a very active area of experimentation in the early part of the twentieth century, beginning with the pioneering research of Marie and Pierre Curie; their isolation of radium and polonium provided a rich source of α particles that could be used to investigate nuclear transmutations. The production of some appropriately long lived radioactive elements by nuclear transmutation has brought us the benefits of nuclear medicine.

There exist three forms of radioactivity: alpha, beta and gamma. Alpha radiation occurs when the nucleus is strongly unstable, releasing a helium nucleus as it transmutes to another lower mass nucleus. Beta radioactivity arises when a neutron in the nucleus transmutes into a proton by releasing an electron (plus a neutrino) with the consequent release of energy. This incidentally proves that neutrons, being heavier than protons, are unstable on their own, with a half life of about 10 minutes. Gamma radiation arises when the daughter nucleus is born in an excited nuclear state and decays into a lower energy level. With respect to beta decay, Fermi formulated this nucle-onic change of character with a famous model which I shall describe later. Some fifty years afterwards, the Fermi model underwent fur-ther modification: it was shown that beta decay was actually due to a short-range weak force mediated by heavy weak bosons and it earned several physicists Nobel Prizes. We will revisit this topic anon.

Chapter 2

Force Fields

The idea of a force field between two interacting bodies grew from the law of conservation of energy and the concept of a *potential*. The notion of energy is ubiquitous today and is taught from an early age, but we must realise that it had a long gestation period before it achieved its present preeminent status. Nowadays we recognise many different forms of energy (all with intrinsic ability to do work): mechanical, heat, electrical, chemical etc. It is an article of scientific gospel that energy can never be created nor destroyed but only converted from one form into another. It owes its modern formulation to Mayer and Helmholtz. Eventually, when Einstein established the connection of rest energy E to mass m through his famous equation $E = mc^2$, all the quantitative measurements in chemistry and atomic and nuclear physics were transparently explained.

Let us trace the notion of force fields in its various forms from its humble beginnings. . .

2.1 Gravitational fields

When things drop to Earth from a height h they gain mechanical energy of motion. By the mid 1800s or even much earlier it became utterly clear that their final speeds v were related to h through the relation $v^2 = 2gh$ where g is the gravitational acceleration. From this arose the concept of *kinetic* energy $mv^2/2$ of a mass m — which had been converted from the drop in *potential* energy mgh. A picture came into view of a gravitational potential $V(r) = G_N M/r$ at

Fig. 2.1   Julius Robert von Mayer.

Fig. 2.2   Hermann von Helmholtz.

a distance r from a point source and G_N is the Newtonian constant. Equipotentials can be viewed as concentric spheres surrounding the source. Radial normals to those spheres represent the gravitational force 'fields' emanating from the centre; the equipotential gradients $g = -dV/dr = G_N M/r^2$ then determine the magnitudes of gravitationally attractive acceleration to the source. To make an optical analogy, the fields can be imagined as 'rays' and the equipotentials as the 'wavefronts' of light. See Figure 2.3. Defining gravitational *flux* as the product of the area through which the fields pass and the fields themselves, namely $4\pi r^2 \times G_N M/r^2 = 4\pi G_N M$, we see that it is a measure of the mass enclosed. This is a general Gauss-like theorem and applies in fact to any mass distribution and any surrounding closed surface. The electromagnetic analogy is more widely known and we shall come to that presently.

Gravimeters are highly sensitive accelerometers which measure the local gravitational fields. On Earth they are used in geodesy and mineral exploration to detect variations in the local mass distributions. Thus they are of great value to prospectors and seismic researchers.

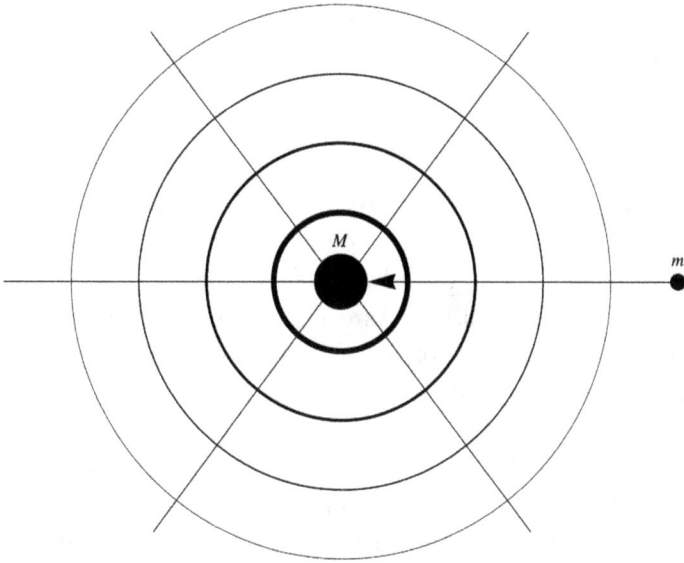

Fig. 2.3 Gravitational fields (radially in/out) and equipotentials (spheres) associated with a point mass. The magnitude of the gravitational force is given by the potential gradient.

2.2 Electromagnetic fields

What we have described for gravity readily transcribes to electrostatics since Coulomb's law of repulsive/attractive force is also of the inverse square type; for two electric charges Q and q one finds that the Coulomb force is

$$F \propto Qq/r^2,$$

up to a proportionality factor that depends on the system of units used. (For the SI system of units the factor is $1/4\pi\epsilon_0$ where ϵ_0 is the permittivity of free space.) Thus we can envisage an electric field $E \propto Q/r^2$ emanating from a source charge Q and acting on a test charge q, like in Figure 2.3. The equipotentials then correspond to equal voltage surfaces and the E acts normally to those surfaces. The electric flux passing through a sphere surrounding a point charge Q is proportional to $4\pi r^2 \times Q/r^2 = 4\pi Q$. With more general configurations this result remains true and Gauss theorem informs us that electric flux passing through any closed surface measures the electric charge Q contained within it. This is summarised by the first of the four Maxwell equations which we are coming to.

Magnetostatics differs from electrostatics in as much as the simplest magnetic fields B are of the dipole type, never of the monopole type as one gets from an isolated charge. While one can coax the two opposite charges in an electric dipole to separate, if one tries the same thing for magnets, one always ends up with a pair of magnets. (See Figures 2.12 and 2.13.) As every schoolchild knows, a picture of magnetic dipole fields of force is instantly uncovered by sprinkling iron filing in the vicinity of a magnet. One might conceptualise surfaces normal to these lines of force as serving the role of magnetic potentials; however the reality is subtler than that! Magnetic fields are actually produced by circulating charges. At a microscopic level the spins of nuclei and orbiting electrons create atomic magnetism; their collective cooperation in a solid then creates magnetic domains and thus magnets. At a macroscopic level circulating currents in the Earth's iron core produce what seem to be the Earth's magnetic N and S 'poles', an effect which is more pronounced in larger planets like Jupiter. (For that reason the magnetic axis does not quite align with the geographic N–S axis.) Because magnetic monopoles do not exist — and this is a real physical conundrum — the Gauss law for magnetism states that the magnetic flux passing through a fully closed surface sums to zero: basically there is as much magnetic flux coming in as going out. This is embodied in the second of four Maxwell equations relating to the magnetic field alone.

When magnets and charges are set in motion a whole new panorama comes into view. Moving charges of course correspond to electric currents. Ampère and Oersted established that a straight wire carrying current J produces a magnetic field B circulating round the wire which falls off inversely with distance r from the wire: $B \propto J/r$; furthermore this field serves to attract another parallel wire. The converse effect was established by Faraday and Lenz: they noticed that a magnetic field can induce a current in a wire whose motion

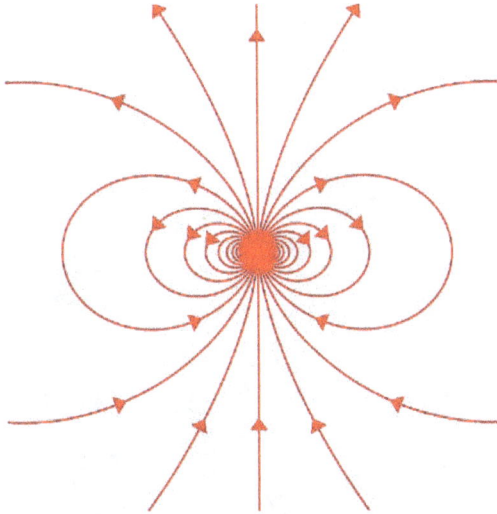

Fig. 2.4   Magnetic field produced by a magnetic dipole or bar magnet.

cuts through the magnetic field flux. Little would the discoverers of these fundamental electromagnetic phenomena have contemplated how these effects would prove to be the entire basis for the modern electrical power and transport industries!

Dynamos or electrical power generators supply alternating current (AC) in their simplest form but require a 'commutator' to convert that into direct current (DC); alternatively AC can undergo 'rectification' to produce DC. The generators are based on Faraday's law of induction whereby a conducting coil placed in a magnetic field which is *forced to rotate* mechanically will generate an electromotive force and a resulting fluctuating current.

Conversely, an electric motor converts electricity into mechanical motion: copper coils are placed in a strong magnetic field and brushes serve to commutate the torque on the coils into unidirectional rotation. It is reported that Faraday had clear visions of electromagnetic fields but the mathematical expression of their variations had to await Maxwell's superior knowledge of calculus needed to formulate his famous equations.

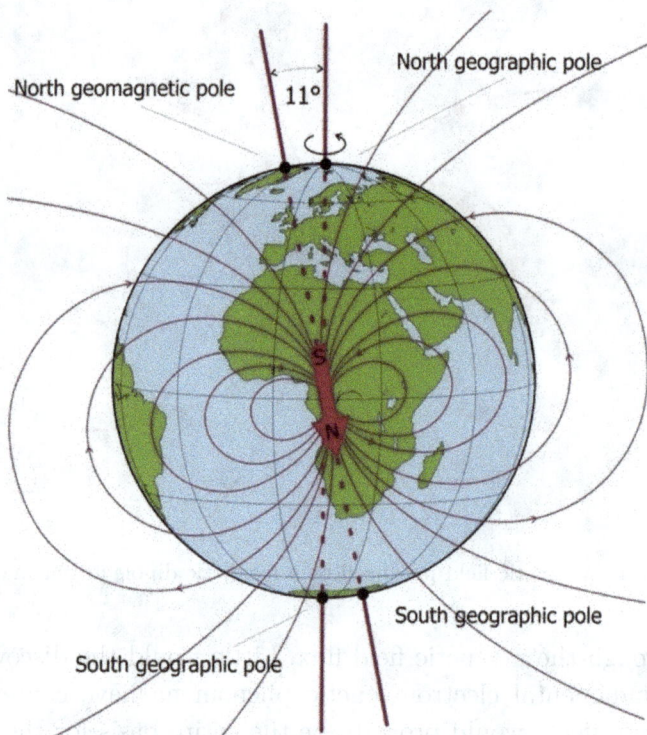

Fig. 2.5 Magnetic dipole analogy for Earth caused by circulating iron convection currents in Earth's interior.

2.3 Strong nuclear fields

The protons are so strongly bound to neutrons in atomic nuclei that it is obvious the nuclear force is short range and serves to make the nuclei 100,000 smaller than the surrounding cloud of electrons. After the discovery of pi mesons, whose existence was predicted by Yukawa, a simple explanation for the small scale structure was established. In fact for the larger nuclei there is such an intimate mixture of nucleons that altogether they behave like a fluid. So it was that the liquid drop model was invented by Weizsäcker which provided a semiempirical formula for the energy of a nucleus based on its atomic mass A and atomic number Z.

For a more detailed picture of the nuclear binding energies and for understanding the stability of certain nuclei, a nuclear shell model was devised by Jensen and Goeppert-Mayer, in analogy to atomic theory, and it enjoys much success, such as explaining the improved stability of magic number nuclei. Experiments on nuclear reactions continue to provide more sophisticated elaborations of nuclear force fields and proton accelerators play a crucial role in these investigations. Indeed particle physics grew out of the activities of ever more powerful atom smashers for which Ernest Lawrence (starting with the cyclotron) deserves great credit. A plethora of new particles were discovered in this way and there was considerable confusion about their nature until the the quark model of Gell-Mann and Zweig [1] put some semblance of order into that chaos.

Nowadays the basic nucleons are believed to consist of three quarks (coming in three colours) held together by a more basic strong force, while mesons like pions consist of a quark strongly attached to an antiquark which resist separation. The study of this type of

Fig. 2.6   J. Hans D. Jensen.   Fig. 2.7   Maria Goeppert-Mayer.

fundamental force is known as quantum chromodynamics (QCD). Mesons and protons are colourless or white with the constituent quarks' colours 'cancelling' each other; this phenomenon is termed 'infrared slavery'. Interestingly the confirmation of the quark model first came from high energy scattering of electrons from nuclei, not from proton scattering. It turns out that QCD force fields acting on the quarks themselves are similar to the quantum electrodynamics (QED) force field which holds atoms together; both are vectorial in character as we shall see but with one *huge* difference: whereas electromagnetic force decreases rapidly with distance from the electrical source, the chromodynamic force *increases* very rapidly with separation from the colour source; so much so that when pulling a quark out of a nucleus, at a critical separation the vacuum splits into a quark–antiquark pair. This is similar to trying to separate a magnet into two magnetic monopoles — it can never happen. Likewise one cannot isolate a coloured particle, such as a quark, however hard one tries, as depicted in Figure 2.14.

Fig. 2.8 Carl Friedrich von Weizsäcker.

Fig. 2.9 Wolfgang Pauli.

2.4 Weak decay fields

We remind the reader that most particles, such as electrons, protons and neutrons carry intrinsic angular momentum called spin. This spin of an electron say is quantised into two $\hbar/2$ states, *up* and *down*: when the spin is positive along the direction of motion the electron is said to carry right-handed helicity or has 'positive chirality'; when the spin component is opposite to the direction of motion it carries left-handed helicity or is in a 'negative chirality state'. The same notion applies to photons of light where the terminology left and right polarisation is used but the spin components are $\pm\hbar$. A reflection in a mirror changes right polarisation into left polarisation as one can easily check for oneself by placing a right corkscrew beside a mirror and examining closely the reflection. All three forces and fields considered previously are blind to chirality in as much they act equally on right and left systems. Not so with weak interactions!

In beta decay, whereby a neutron in a compound nucleus can disintegrate into a proton and an electron (plus a neutrino), we come across significant differences between the weak mechanism and the previous forces. These took a long time to disentangle. First of all there was an *apparent* contradiction with the principle of energy and angular momentum conservation until it was resolved by Pauli by invoking a new shadowy particle called the neutrino which carried the excess angular momentum and energy. Secondly the spontaneous disintegration of the neutron seemed to occur as a point-like interaction between the participating particles without requiring a intermediary field. Thirdly, and most shocking of all, the beta decay process seemed to violate mirror symmetry — the emanating neutrino appeared to be always left-handed, no matter the decay process. This means that the weak decay mechanism leading to beta decay is *not* blind to chirality and offers an absolute way of distinguishing left from right.

Fermi developed a relativistic model for such point interactions between currents but left open the Lorentz tensor structure of the contact exchanges. It had to be contrasted with QED where for example the electron–proton attraction is not instantaneous but

Fig. 2.10   Murray Gell-Mann.   Fig. 2.11   George Zweig.

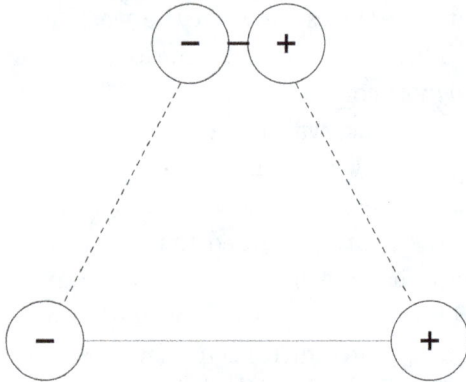

Fig. 2.12   Electric dipole charge separation into two charge monopoles.

involves photon exchange. See Figure 2.15 for a diagrammatic explanation and its algebraic expression.

Based on the assumption that neutrinos were massless (through the principle of chirality invariance) and travel at the speed of light,

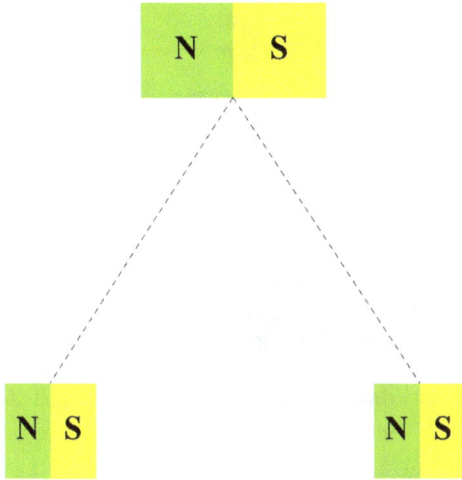

Fig. 2.13 Magnetic dipole separation into two magnetic dipoles. Magnetic monopoles do not exist!

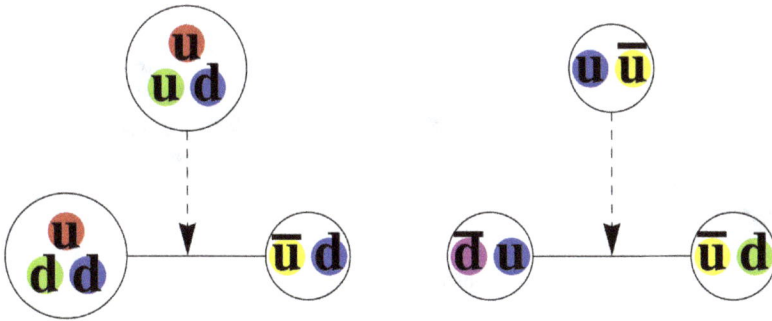

Fig. 2.14 Quark separation as in the processes (virtual) $p \to n\,\pi^+$ (on left) and ρ^0 meson $\to \pi^+\pi^-$ (on right). The vacuum energy needed to separate the two quarks goes into creating a quark–antiquark pair.

Salam [2] proposed that the neutrinos were so-called massless Weyl fermions and came in only one chirality state, namely left-handed. Lee and Yang [3] proposed testing mirror symmetry breaking experimentally for various weak processes and were duly rewarded observationally, the first such experiment being performed by Wu [4]. Even after this surprising discovery of parity breaking there remained

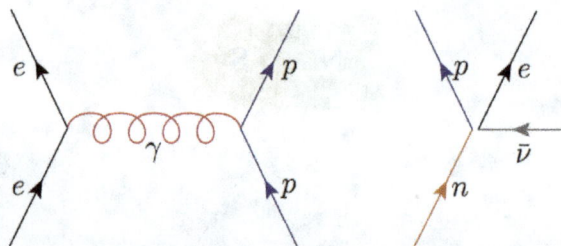

Fig. 2.15 The $e - p$ force (diagram on left) is mediated by γ exchange and its amplitude is calculated by working out $J_e \, \Delta_\gamma \, J_p$ where Δ_γ is the photon propagator and J_x is the current carried by the motion of particle x. It is possible to break up the current into a chiral right and chiral left, so $J = J_R + J_L$. In the analogous Fermi model (drawn on right) one calculates just $J_{nL} \, J_{eL}$ instead and ignores intermediate meson propagation.

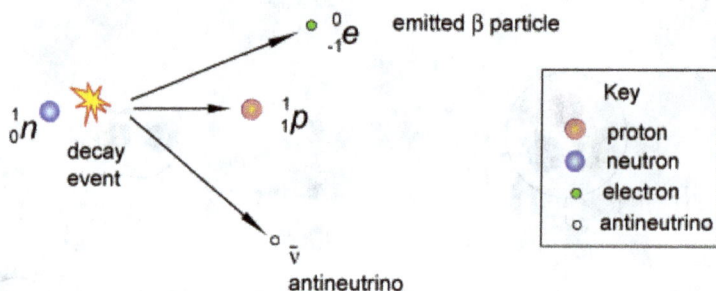

Fig. 2.16 Original contact picture of instantaneous β-decay.

a period when the precise form of the Fermi contact interaction remained murky. It required further experiments for the form to be pinpointed: namely in the form of a (V–A) interaction which favours left-handedness. Marshak and Sudarshan [5] and Feynman and Gell-Mann [6] gave the theoretical basis for such a vectorial weak structure and its connection with Fierz reshuffling.

A decade or more was to pass before the idea of a weak contact interaction was itself overturned. That is a fascinating story which I shall come to later and is described mathematically in the Appendix. We now know that there is indeed a weak field

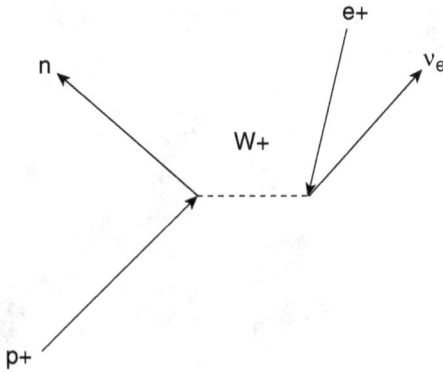

Fig. 2.17   Modern view of W-mediated β-decay. This is much closer to the $e - p$ process shown in Figure 2.15.

Fig. 2.18   Tsung-Dao Lee.

Fig. 2.19   Chen-Ning Yang.

acting between participant particles; in the case of beta decay it is the so-called W-meson which gets exchanged. (See Figures 2.16 and 2.17.) To complete the picture, it turns out that there is another weak boson, named the Z^0 which carries no charge but

Fig. 2.20   Chien-Shiung Wu.

Fig. 2.21   Abdus Salam.

Fig. 2.22   Markus Fierz.

Fig. 2.23   Richard Phillips Feynman.

Fig. 2.24   Robert Eugene Marshak.

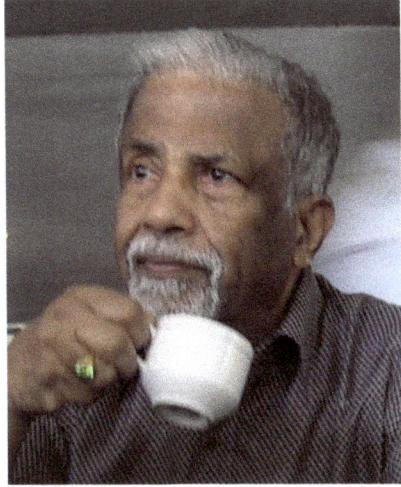

Fig. 2.25   E.C. George Sudarshan.

is able to interact with the ghostly neutrinos and other leptons (like electrons and muons) without changing their character, apart from altering their motion. This purely neutral weak interaction was discovered experimentally at CERN in a huge bubble chamber, called Gargamelle. While on this topic I should mention that the Z^0 meson lifetime is important in serving to determine that the number of active light neutrinos is limited to three, in agreement with observation.

Chapter 3

Field Disturbances

In this chapter we will try to explain how disturbances in fields can be resolved into components (or "modes") and how they propagate; thus we cannot avoid injecting some mathematics into the discussion. More intricate details about such decompositions have been hived off to the Appendix for readers with appropriate mathematical background. I shall start with a simple example which illustrates a lot of points and progress from there to the four basic force fields. Inevitably I will be obliged to introduce some formalism but most of it will be elementary trigonometry.

Imagine a violin string under tension clamped at each end. A violinist wants to create a pizzicato note by plucking it at the centre, producing the tent-shaped profile of Figure 3.1. The string has length L in some units and the midpoint pluck at $x = 0$ has unit magnitude say. The profile pluck function therefore can be expressed in the form:

$$\phi(x) = \begin{cases} 1 + 2x/L; & -L/2 \le x \le 0 \\ 1 - 2x/L; & 0 \le x \le L/2. \end{cases} \tag{3.1}$$

Now, in that interval, this function can be resolved into combinations of a fundamental cosine wave $\cos(\pi x/L)$ and its overtones with all terms vanishing at the endpoints $x = \pm L/2$. This kind of expansion is called a Fourier series and written as

$$\phi(x) = \sum_{n=1}^{\infty} \varphi_n \cos(n\pi x/L), \tag{3.2}$$

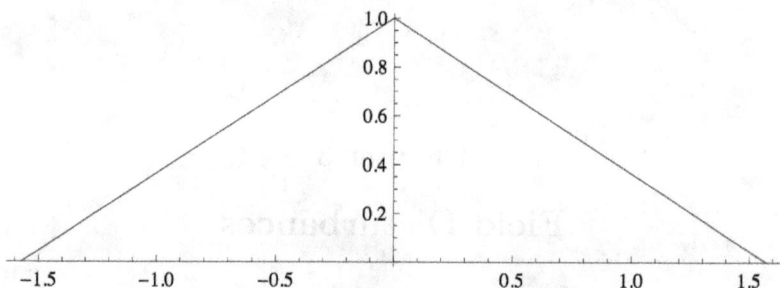

Fig. 3.1   A violin string of length $L = \pi$ that is plucked initially at the centre.

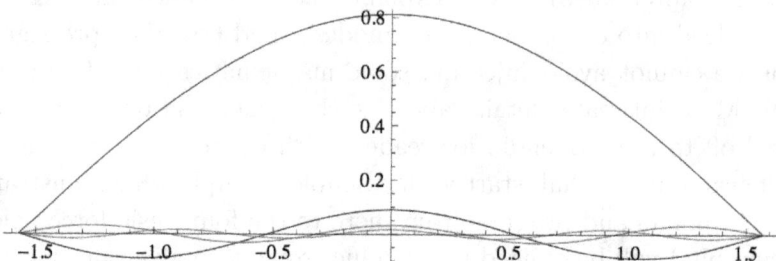

Fig. 3.2   The sinusoidal wave and its first few overtones (and corresponding amplitudes) which make up the initial profile.

where φ_n represents the 'amplitude' of the nth 'mode'; conversely φ_n can be extracted via the integral

$$\varphi_n = \frac{2}{L} \int_{-L/2}^{L/2} \cos\left(\frac{n\pi x}{L}\right) \phi(x)\, dx = \frac{4}{n^2 \pi^2}(1 - (-1)^n), \quad n \geq 1,$$

(3.3)

owing to the orthogonality properties of the various sinusoids. See Figure 3.2 for the individual overtones and Figure 3.3 for the approximation using the sum of the first few. For more details, see the Appendix.

The energy U stored by the pluck and subsequently released is given by the *square* of the profile, integrated over the entire string, namely

$$U = \int_{-L/2}^{L/2} (\phi(x))^2\, dx = 2 \int_0^{L/2} (1 - 2x/L)^2\, dx = L/3. \qquad (3.4)$$

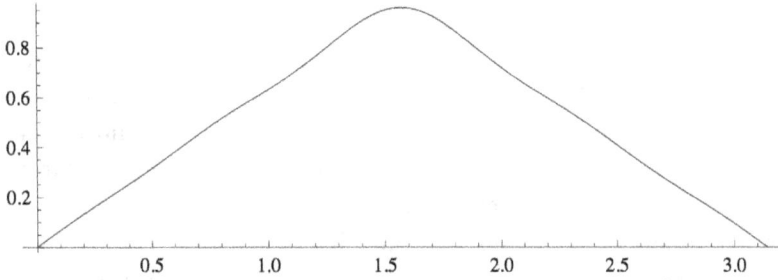

Fig. 3.3 An approximation to the plucked profile using the first four overtones and corresponding terms in the Fourier series.

But because of the orthogonality of cosines it can also be evaluated as the sum of the squares of the mode amplitudes over the string length:

$$U = \sum_n L\,\varphi_n^2/2 = 32L \sum_{n\ \mathrm{odd}} 1/(\pi n)^4. \tag{3.5}$$

Comparing (3.4) and (3.5) it incidentally offers a series method for calculating π that converges rapidly:

$$\pi^4 = \sum_{n=1,3,5,\dots}^{\infty} 96/n^4.$$

That pizzicato profile is of course instantaneous. The string will vibrate in time with some frequency $\nu \equiv \omega/2\pi$ that depends on the string tension and mass and it can be described by the time-dependent profile $\Phi(x,t) = \phi(x)\cos(\omega t)$. This means that for each mode n we encounter the combination $\varphi_n \cos(n\pi x/L)\cos(\omega t)$; we can interpret this as two waves travelling in opposite directions, since there exists the trigonometric identity ($k_n \equiv n\pi/L$)

$$2\cos(k_n x)\cos(\omega_n t) = \cos(k_n x - \omega_n t) + \cos(k_n + \omega_n t).$$

The phase speed of the nth mode can be read off as the ratio $v = \omega_n/k_n$. In fact each of these waves, as well as their combination, obeys the *differential equation for wave propagation* in the x-direction:

$$\frac{\partial^2 \Phi}{\partial x^2} = \frac{k^2}{\omega^2}\frac{\partial^2 \Phi}{\partial t^2}, \tag{3.6}$$

which is of second order in space as well as time. Equation (3.6) is often taken as the starting point for a discussion of wave motion and is normally subject to two appropriate boundary conditions.

All of this can be generalised to an infinite string and any well-behaved profile. (In fact the formalism becomes much neater by making use of complex variables. See the Appendix to understand how this is done.) Starting with any instantaneous continuous profile $\phi(x)$ over x we can now break it up into a *continuous* integral over a continuum of modes k, rather than a discrete sum as in (3.2) and (3.3):

$$\phi(x) = \int_{-\infty}^{\infty} \exp(ikx)\,\varphi(k)\,dk/2\pi \quad \text{with } \varphi(k)$$

$$= \int_{-\infty}^{\infty} \exp(-ikx)\,\phi(x)\,dx. \tag{3.7}$$

Further we can include any sort of time dependence in the profile

$$\Phi(x,t) = \int_{-\infty}^{\infty} \exp[i(kx - \omega t)]\,\varphi(k,\omega)\,dk\,d\omega/(2\pi)^2 \tag{3.8}$$

$$\varphi(k,\omega) = \int_{-\infty}^{\infty} \exp[-i(kx - \omega t)]\,\phi(x,t)\,dx\,dt. \tag{3.9}$$

It turns out that the 'energy' in the profile is propagated forward by the 'group velocity' $V = d\omega/dk$ rather than the individual 'phase velocity' of any component $v = \omega/k$. The sort of disturbance and the nature of the medium will influence how the circular frequency ω depends on the wavenumber k. Only when ω is a linear function of k do all the waves move with the same speed; otherwise the disturbance undergoes 'dispersion'. We shall make use of these formulae at various stages in the discussion.

3.1   Gravitational shocks

Einstein's general theory taught us that the fabric of spacetime is affected by the mass distributed within it. In the limit that there is no mass the fabric is normally 'flat'. It follows that if masses are present and disturbed, so will the fabric be: in general this will produce a superposition of sinusoidal 'gravitational' waves. But all such waves are observationally very hard to detect because their effect on our environment is so very weak, *unless the masses are truly enormous and the apparatus is super sensitive.*

The first breakthrough in anticipating gravitational wave detection came indirectly. A bit of history first. Pulsars were accidentally discovered in 1968 by Bell and Hewish [7] and it was soon established that they corresponded to rotating neutron stars with intense magnetic fields that radiated conical pulses of radio waves (akin to a lighthouse effect) as they arrived on Earth. The times of arrival of these radio signals were measured with exquisite accuracy by atomic clocks — rivalling the accuracy of quantum electrodynamics — and were subject to occasional period glitches, interpreted as starquakes.

It was a fortunate circumstance when Hulse and Taylor [8], at the Arecibo Radio Observatory, discovered a binary pulsar whose radio emissions they measured very carefully. (Only one of the pair was detected to be pulsing, the companion being a 'quiet' neutron star.) Punctiliously monitoring the times of arrival of the pulses allowed accurate determination of the pair's elliptical orbits round their common centre of mass. But more exciting was the fact their orbital period *decreased* with time; they attributed the decrease in energy to emission of gravitational waves whose rate accorded precisely with Einstein's quadrupole radiation formula. This was the first indirect but persuasive evidence for gravity waves.

These observations proved to be the impetus for others to attempt direct detection of the waves. After failed attempts by Weber and others to detect gravitational waves with a resonant cylindrical bar, others have tried a different method through disturbances in path length of the arms of a highly accurate laser interferometer. And the only convincing way to persuade other scientists that the effect was real was to detect identical disturbances in a twin interferometer far away. Imagine the world's excitement in 2016 when two independent observatories (in Washington and in Louisiana) found the same wave interference profiles [9]; they attributed those profiles of gravitational emission to the merger of two rotating black holes. Since that seminal

Fig. 3.4   Jocelyn Bell Burnell.

Fig. 3.5   Antony Hewish.

Fig. 3.6   Russell A. Hulse.

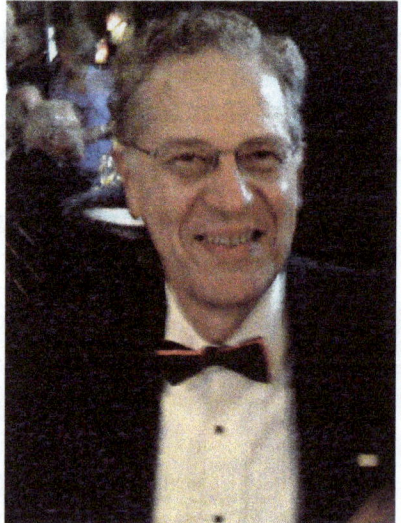

Fig. 3.7   Joseph Hooton Taylor Jr.

discovery, gravity waves have been detected from the collisions of two neutron stars. A new window on astronomical phenomena has thus opened up, distinct from electromagnetic and neutrino astronomy.

3.2 Electromagnetic waves

Although the physical effects of electromagnetism were well under-stood by about 1850, it took the genius of Maxwell to express them mathematically in a form where they could be readily quantified and manipulated. There are just four Maxwell equations involving the electric and magnetic fields, the medium and the sources which pro-duce the fields. I will write them down first before explaining them in physical terms through the mathematical symbolism:

$$(\partial/\partial \mathbf{r}) \cdot \mathbf{D} = \rho \qquad (3.10)$$

$$(\partial/\partial \mathbf{r}) \cdot \mathbf{B} = 0 \qquad (3.11)$$

$$(\partial/\partial \mathbf{r}) \wedge \mathbf{E} + \partial \mathbf{B}/\partial t = 0 \qquad (3.12)$$

$$(\partial/\partial \mathbf{r}) \wedge \mathbf{H} - \partial \mathbf{D}/\partial t = \mathbf{j}. \qquad (3.13)$$

First the symbols themselves:

- The vector differential operator $(\partial/\partial \mathbf{r})$ with Cartesian components $(\partial/\partial x, \partial/\partial y, \partial/\partial z)$ is often written as ∇ in many texts. It repre-sents the rates of change along three orthogonal spatial directions as partial derivatives.
- The 'displacement' vector $\mathbf{D} \equiv \epsilon \mathbf{E}$ depends on the electrical 'permittivity' ϵ of the medium and the applied electric field $\mathbf{E}$.
- Likewise the 'magnetic induction' vector $\mathbf{B} = \mu \mathbf{H}$ is propor-tional to the 'permeability' μ of the medium and applied magnetic field $\mathbf{H}$. Both μ and ϵ contain the response of the medium to the applied fields.
- The cross or wedge product $\wedge$ between two vectorial quantities $\mathbf{u}$ and $\mathbf{v}$ is orthogonal to both and $|\mathbf{u} \wedge \mathbf{v}|$ equals the area of the

parallelogram spanned by the 2 vectors. The wedge product is also vectorial in nature and the differential operation $(\partial/\partial \mathbf{r}) \wedge$ is called taking the 'curl'.

Next the meaning of each of the four equations:

- Equation (3.10) tells us that the electric field emanation or its 'divergence' is provided by the density of charge ρ. No charge, no electric field.
- Equation (3.11) says the same for the divergence of magnetic field, which is always zero because no magnetic monopoles exist; it points to something deeper about the nature of electromagnetism. $\mathbf{D}$ differs from $\mathbf{B}$ in another respect: the former changes sign under spatial reflection or 'space parity' as befits a true 'polar' vector, whereas the latter does not change under reflection; it connotes an 'axial' or screw vector. [The reader is invited to check that all four Maxwell equations behave uniformly under the parity operation.]
- Equation (3.12) tells us that a changing magnetic field will induce an electric field and forms the basis of electricity generation. Faraday was the first to stress this electromagnetic feature and Lenz produced an integral version of (3.12).
- Equation (3.13) is in a sense the dual of (3.12); it states that a time-dependent electric field will induce a magnetic field, which also depends on the current density $\mathbf{j}$, according to Ampère's law.

In the international system of units (SI), which is based on macroscopic units, such as volt, ampere, henry, etc., even the vacuum is accorded a non-unit permeability $\mu_0 = 4\pi \times 10^{-7}$ and permittivity $\epsilon_0 \equiv 1/\mu_0 c^2$. Actually the relation $\epsilon \mu c^2 = 1$, where c is the speed of light in the medium, is regarded as 'being wise after the event'. Let us understand why.

Consider the four Maxwell equations in vacuum (subscript 0) in the absence of sources, so $\rho = \mathbf{j} \to 0$:

$$(\partial/\partial \mathbf{r}) \cdot \mathbf{E} = 0, \quad (\partial/\partial \mathbf{r}) \cdot \mathbf{H} = 0 \tag{3.14}$$

$$(\partial/\partial \mathbf{r}) \wedge \mathbf{E} + \mu_0 \partial \mathbf{H}/\partial t = 0, \quad (\partial/\partial \mathbf{r}) \wedge \mathbf{H} - \epsilon_0 \partial \mathbf{E}/\partial t = 0. \tag{3.15}$$

Taking the 'curls' of (3.15), using the vectorial identity

$$\partial/\partial \mathbf{r} \wedge (\partial/\partial \mathbf{r} \wedge \mathbf{V}) = \partial/\partial \mathbf{r}(\partial/\partial \mathbf{r} \cdot \mathbf{V}) - (\partial/\partial \mathbf{r} \cdot \partial/\partial \mathbf{r})\mathbf{V}$$

and applying (3.14), we find that both electric and magnetic fields obey

$$\triangle \mathbf{E} \equiv \left(\frac{\partial}{\partial \mathbf{r}} \cdot \frac{\partial}{\partial \mathbf{r}}\right)\mathbf{E} = \epsilon_0\mu_0\frac{\partial^2}{\partial t^2}\mathbf{E}, \quad \triangle \mathbf{H} \equiv \left(\frac{\partial}{\partial \mathbf{r}} \cdot \frac{\partial}{\partial \mathbf{r}}\right)\mathbf{H} = \epsilon_0\mu_0\frac{\partial^2}{\partial t^2}\mathbf{H}.$$

$$(3.16)$$

These are three-dimensional wave equations corresponding to a wave speed $1/\sqrt{\epsilon_0\mu_0}$. These calculations were originally carried out by Maxwell in the old CGS system of units, where the values of permittivity and permeability were observationally known. It must have been a fabulous *eureka* moment to savour when Maxwell discovered that the wave speed coincided with the speed of light, which had already been measured by Roemer, Michelson and Fizeau. The penny finally dropped that light corresponds to electromagnetic waves! Furthermore (3.15) tells us that the electric and magnetic fields are orthogonal to one another and to the direction of wave propagation, explaining why light has two linear polarisation modes (or equivalently circular polarisations).

Most of our knowledge about the universe and its evolution rests on observations of electromagnetic waves received on Earth from outer space. Initially astronomy developed from visible light receptors as they were the most obvious radiations that penetrated the atmosphere; only relatively recently did they extend to radio waves and gamma rays which can be probed too. But it is only in the last century that humankind has greatly widened the spectrum of observations through satellite, balloon or mountain-top observatories where our atmospheric shield becomes negligible. We now have good data in ultraviolet, infrared, X-ray regions and a better understanding of stars and galaxies from those parts of the electromagnetic spectrum. We also know how to generate these parts of the spectrum quite routinely so as to make use of them: witness the explosion of radio communication which has completely transformed our societies as have X-rays for medicine.

3.3   Strong signals

The fact that chromodynamic forces are so strong and confining [10] (because they grow with distance) means that at the most fundamental level strong interactions are very hard to notice. The gluons mediating the colour force are trapped within the confines of nuclei like the quarks; however there is indirect confirmation of their existence through the secondary jets — generally two or three in number — that are observed in high energy collisions where showers of colourless particles are produced. We will have to be content with that evidence.

At a less primordial level, one can certainly detect strong fields in the form of meson showers that are produced in cosmic rays. They are primarily instigated by very high energy cosmic protons colliding with atmospheric atoms; they then get accompanied by secondary gamma rays — of very particular energy as they arise from π^0 decays — as well as muons that are produced from $\pi^\pm$ decays. Alpha radiation coming from decays of unstable long-lived nuclei are undoubtedly strong and quite destructive, but that can occasionally be of benefit in medical treatments. Anyhow it is true to state that forces associated with α sources are not considered to be as fundamental as colour forces.

3.4   Weak field transmutation

This sort of radiation is also hard to get hold of but mainly for a different reason. The mediating weak particles that are responsible for beta-type decays do not go very far at ordinary energies: they are 'short-range'. This is in contrast to the products of those interactions which do go far indeed without change as is the case with neutrinos! These secondary emissions are prevalent from radioactive sources and are also of great use in medicine. In particular there are some short-lived beta emitters which can produce positrons; these can then annihilate against ordinary electronic matter and produce two oppositely directed gamma rays of about $500 \, \text{keV}$, pinpointing the precise seat of the annihilation itself — the location where the treatment is supposed to be taking effect. That is the essence of the PET scan.

Other products of cosmic rays are muons (basically, heavy electron twins) which are measurably long-lived because of their high speeds; they only interact weakly and electromagnetically with matter and are more useful than electrons because of their larger mass. Finally there are the neutrino emissions which pass through matter with hardly any interactions — which is why they took so long to detect. We only see neutrinos if they are produced in large quantities, such as from nuclear reactors or stars, or if the detecting equipment is extremely sensitive to their rare disturbance of atoms. More about that later.

Chapter 4

Enter the Relativistic Quantum

Quantum theory was launched at the turn of the twentieth century by Planck's introduction of the fundamental unit $h \equiv 2\pi\hbar$. One of the outstanding puzzles at that time was why classical physics failed to explain the spectrum of a hot blackbody despite accurate measurement of its characteristics. Only at long wavelengths was there any resemblance to the classical Rayleigh–Jeans law, but at higher frequencies the classical predictions failed miserably and, worse still, led to an unacceptable infinite value for the total energy of a radiant body.

To set the scene consider simple harmonic vibrations at a particular frequency ν as might exist in a cavity. In classical physics the amplitude or energy U of any mode is a continuous variable and can in principle range from 0 to ∞ depending on how large the amplitude is. Placed in an enclosure at temperature T, statistical mechanics tells us that the probability for a cavity oscillator to have energy U is given by the Maxwell–Boltzmann factor[1] $\exp(-U/\ell T)$, half of U being kinetic and half potential energy. The average energy in an ensemble of such oscillators per oscillator mode is therefore given by the *continuous* integral

$$\langle E \rangle = \frac{\int_0^\infty U \, \exp(-U/\ell T) \, dU}{\int_0^\infty \exp(-U/\ell T) \, dU} = \ell T, \tag{4.1}$$

in agreement with the classical equipartition theorem.

---

[1] ℓ is Boltzmann's constant and it acts a bridge between energy and the Kelvin temperature scale.

However Planck pointed out that if the possible energies were discretised at values $U_n = nh\nu \equiv n\hbar\omega$ then Eq. (4.1) becomes modified to a sum rather than an integral and the mathematics yields:

$$\langle E \rangle = \frac{\sum_{n=0}^{\infty} nh\nu \exp(-nh\nu/\ell T)}{\sum_{n=0}^{\infty} \exp(-nh\nu/\ell T)} = \frac{h\nu}{\exp(h\nu/\ell T) - 1}, \qquad (4.2)$$

which only coincides with Eq. (4.1) when ℓT is much larger than the energy level spacing of $h\nu$. (This is a general result which applies to all systems, even when the energy levels are not equidistant from one another.) The rest of the argument follows standard counting of vibrational modes per unit volume of blackbody container. In the frequency interval ν to $\nu + d\nu$ it results in Planck's famous formula for radiant spectral energy density distribution:

$$dE = \frac{2\pi h\nu^3/c^2}{\exp(h\nu/\ell T) - 1} \, d\nu. \qquad (4.3)$$

This provides a perfect fit at all frequencies, with all its consequences. Thus began quantum theory.

4.1   Enter the quantum

One of the principal founders of the old quantum theory was Einstein himself. He embraced the quantum idea fully by interpreting the photoelectric effect correctly (for which he was awarded the Nobel Prize) and stipulating that light of frequency ν itself consists of quantum particles called *photons*, carrying energy units $h\nu$. In some sense he was reviving Newton's corpuscular idea of light, except that he could not avoid reconciling the concept with the well-established wave theory of light. This had to await the full development of quantum theory and a move to a more abstract 'Hilbert space' involving states and operators. Nowadays we regard a beam of light as consisting of a stream of 'wave packets' possessing some distribution of light frequencies. Each packet corresponds to a set of photons.

Einstein went much further with quantum ideas. By invoking discrete energy units he was able to explain the observed behaviour of solid and gaseous heat capacities (later improved by Debye). The whole statistical edifice hinges on the replacement of the classical equipartition energy value of ℓT by the quantum corrected value of $h\nu/[\exp(h\nu/kT) - 1]$. The dependence of experimental results on the separation of *quantised* energy levels at various temperatures is therefore the heart of the matter.

Let us return to the example of the harmonic oscillator to see how these unitised energy levels arise. A simple harmonic oscillator of fundamental circular frequency ω is characterised by an energy (or Hamiltonian H) given by equal amounts of kinetic and potential energy

$$H = p^2/2m + m\omega^2 x^2/2 = (p + im\omega x)(p - im\omega x)/2m + \hbar\omega/2.$$
(4.4)

Some readers who are trying to follow the algebra may be perplexed why I have rustled in an extra term $\hbar\omega/2$ on the right of Eq. (4.4). That is because the order of momentum p and displacement x *does matter*; this is intimately connected with the Heisenberg uncertainty principle of quantum mechanics which reminds us that a velocity measurement interferes with a position measurement, encapsulated by the famous 'commutation relation' $xp - px \equiv [x, p] = i\hbar$, which is engraved on Born's tombstone.

Fig. 4.1   Ludwig Boltzmann.

Fig. 4.2   Satyendra Nath Bose.

Fig. 4.3 Max Karl Ernst Ludwig
Planck.

Fig. 4.4   Albert Einstein.

It is useful to define

$$a = \frac{p - im\omega x}{\sqrt{2m\omega\hbar}}, \ a^\dagger = \frac{p + im\omega x}{\sqrt{2m\omega\hbar}}, \quad \text{so } H = \left(a^\dagger a + \frac{1}{2}\right)\hbar\omega. \quad (4.5)$$

The ground state or lowest of the oscillator (with energy $E = \hbar\omega/2$) is 'annihilated' by the action of a, while the excited states are 'created' by successive application of the operator $a^\dagger$. [The ground state energy must be nonzero, otherwise it would imply that there is no motion and the equilibrium position ($x = 0$) is known with absolute precision, contradicting Heisenberg's position–momentum uncertainty relation.] While on this point, notice that $[a, a^\dagger] = 1$ is a dimensionless quantity. This idea of creating and annihilating excitations of energy levels above the ground state is crucial for understanding the quantisation of field disturbances. Figure 4.5 illustrates their operation on the oscillator levels in the parabolic x^2 potential.

If there are several independent oscillators (labelled by i) with frequencies $\hbar\omega_i$ we may construct a series of annihilation and creation

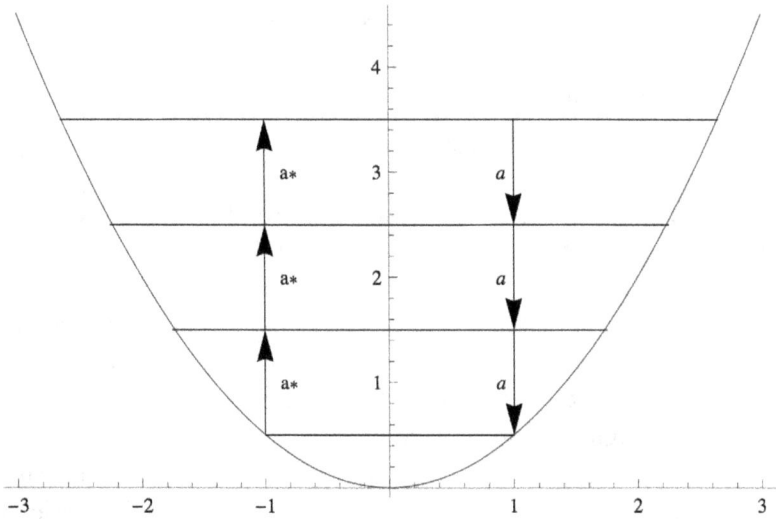

Fig. 4.5 Oscillator energy levels $E_n = (n + 1/2)h\nu$ and the action of creation and annihilation operators.

operators $a_i, a_i^{\dagger}$ obeying the rule[2]

$$[a_i, a_j^{\dagger}] = \delta_{ij}, \; [a_i, a_j] = [a_i^{\dagger}, a_j^{\dagger}] = 0 \quad \text{so that } E = \sum_i (n_i + 1/2)\hbar\omega_i.$$

The actions of the various a_i raised to any power on the vacuum or empty state generate a so-called 'Fock' space of states.

4.2 Elegant relativistic equations

Let us backtrack a few years. It is correct to say that special relativity, born in 1905, was the main stepping stone to gravitation and GR because it introduced the concept of a four-dimensional spacetime and showed that our notions of position $\mathbf{r}$ and time t are inextricably linked. The linkage depends on the motion of (inertial) observers relative to one another and the constancy of the speed of light c.

---

[2] The Kronecker δ_{ij} is defined to be unity for $i = j$, 0 otherwise, and applies to a discrete distribution. Its counterpart for a continuous variable rather than a discrete variable is the Dirac $\delta_{kk'}$ which vanishes for $k \neq k'$ but gives an integrable infinity around $k = k'$ in as much as $\int f(k)\delta(k) \, dk = f(0)$.

Minkowski, Einstein's teacher, tidied up the Lorentz transformation of spacetime coordinates from one inertial observer to the next by expressing the equations for the square of event separation

$$ds^2 \equiv c^2 dt^2 - d\mathbf{r}^2 = c^2 dt^2 - dx^2 - dy^2 - dz^2 \qquad (4.6)$$

in a beautiful 4D framework. A spacetime event is construed as some change of configuration usually involving emission or absorption of a quantum particle. The event separation thereby vanishes for motion of light in vacuo, independently of observer; c remains invariant for any inertial observer and explains why the Michelson–Morley experiment was unable to detect a background ether through which the Earth was thought at the time to move.

Disregarding gravitation, the square of event separation can be written[3] as $ds^2 = dx^m dx^n \eta_{nm}$, where the Minkowski metric η_{nm}

Fig. 4.6   Hendrik Lorentz.   Fig. 4.7   Hermann Minkowski.

---

[3]We are adopting Einstein's convention here. Repeated indices (up and down) are to be summed. Coordinate x^0 corresponds to ct and $\mathbf{r} = (x^1, x^2, x^3) = (x, y, z)$. The inverse metric of empty or 'flat' spacetime η^{mn} has identical elements to η_{mn}.

is a diagonal matrix having nonzero elements $(\eta_{00}, \eta_{11}, \eta_{22}, \eta_{33}) = (1, -1, -1, -1)$. The Lorentz transformations between one coordinate system x to another x' can then be constructed as appropriate four-dimensional 'rotations'. These include 3D space rotations as well as relative velocity transformations between observers. Disturbances $\Phi(x)$ appearing in spacetime are found by acting with the derivative or 4D 'gradient' operator: $\partial_n \equiv \partial/\partial x^n$ with components

$$(\partial_0, \partial_1, \partial_2, \partial_3) = \left(\frac{\partial}{c\partial t}, \frac{\partial}{\partial x}, \frac{\partial}{\partial y}, \frac{\partial}{\partial z} \right).$$

Letting $\partial^m \equiv \eta^{mn} \partial_n$ we can construct the Lorentz invariant d'Alembertian

$$\Box \equiv \partial^n \partial_n = (\partial/c\partial t)^2 - (\partial/\partial \mathbf{r}) \cdot (\partial/\partial \mathbf{r}) = (\partial/c\partial t)^2 - \triangle$$

which takes the same form for every inertial observer. Notice that the wave (3.16) for electric and magnetic fields has precisely this form: $\Box \Phi = 0$, where Φ is $\mathbf{E}$ or $\mathbf{H}$. In fact it was the recognition of the Lorentz invariance of Maxwell's equations (3.10) to (3.13) which convinced Einstein about the truism that the speed of light must be the same for all inertial observers regardless of their relative velocities and led to his abandonment of ethereal notions.

Before we think about quantisation, let us return to electrodynamics and delineate a neat four-dimensional formulation for it will lead us into deeper waters. It neatly packages Maxwell's three-dimensional formulation. The reader is invited to check that if one invokes an antisymmetric 4D matrix, $F_{mn} = -F_{nm}$ with components

$$F_{01} = E_1/c, \quad F_{21} = B_3, \text{ and permutations}, \tag{4.7}$$

and a 4D current $J_0 = \rho/\epsilon c$, $\mathbf{J}_1 = -\mu \mathbf{j}_1$, etc., all four Maxwell equations drop out elegantly from just two equations:

$$\partial^m F_{mn} = J_n, \quad \partial_l F_{mn} + \partial_m F_{nl} + \partial_n F_{lm} = 0. \tag{4.8}$$

This is an indicative of a more basic underpinning.

For that we return to Maxwell's equations in 3D, for the two sourceless equations (not involving charges or currents). Equation (3.11) tells us that the magnetic field is 'curly' or 'solenoidal' and can be neatly expressed in the form $\mathbf{B} = (\partial/\partial\mathbf{r}) \wedge \mathbf{A}$, where $\mathbf{A}$ is called the 'vector potential' which is undetermined up to the gradient of a scalar field. Similarly Eq. (3.12), which then reads

$$(\partial/\partial\mathbf{r}) \wedge (\mathbf{E} + \partial\mathbf{A}/\partial t) = 0,$$

signifies that the combination $\mathbf{E} + \partial\mathbf{A}/\partial t = -\partial\phi/\partial\mathbf{r}$, is the gradient of a 'scalar potential' ϕ, undetermined up to a constant. The 4D packaging becomes complete once we recognise that both scalar and vector potentials form part of a 4D potential: $A^m = (\phi/c, A^1, A^2, A^3)$ with $F_{mn} \equiv \partial_m A_n - \partial_n A_m$ playing a secondary role. It then becomes obvious that Maxwell F_{mn} remains unaffected by the 'gauge change' $A \to A + \partial\theta$.

The whole electrodynamic edifice rests on a *single* equation for the 4D potential:

$$\partial^m(\partial_m A_n - \partial_n A_m) = J_n. \qquad (4.9)$$

It could hardly be more elegant! The only point of vagueness about Eq. (4.9) is that the 4D potential possesses a so-called 'gauge invariance' under the 'gauge transformation' $A_m \to A_m + \partial_m\theta$, and this can be sometimes exploited in certain situations by 'fixing a gauge' for A. This will become necessary when one tries to quantise the scheme; for instance one may settle for the Lorentz or Landau gauge where $\partial^m A_m = 0$ so that the equation of motion reduces just to d'Alembert's $\Box A_n = J_n$.

We will intermittently come across other sorts of fields, which belong to other Lorentz group representations. Denoting a generic field by $\Psi(x)$, it will obey a free field equation of the type $\mathcal{O}(\partial)\Psi(x) = 0$. No matter the form of $\mathcal{O}$ it should lead to the Klein–Gordon (KG) equation $(\Box + m^2)\Psi(x) = 0$ for a particle of mass m, or simply $(p^2 - m^2)\psi(p) = 0$ for the Fourier transform of Ψ as in Eq. (4.11) below.

4.3   Enter quantum fields: Natural units

It became apparent early that disturbances in and of spacetime come in quanta, like photons, and can be identified with particles. This even applies to electrons and nucleons and implies that they can behave as waves in suitable circumstances. Wave packets come in the form of distributions over various frequencies satisfying 'wave equations'. This then becomes the point of contact with quantum theory as we are dealing there with wave functions obeying Schrödinger's equation in the non-relativistic regime (and its relativistic counterpart equation at high energy). Now special relativity stipulates that the relation between energy E and momentum p is given by the relation

$$(E/c)^2 - \mathbf{p}^2 = m^2 c^2 \equiv p^m p_m, \qquad (4.10)$$

where p is the so-called four-momentum with components $(E/c, \mathbf{p})$. The link with quantum mechanics comes through the operational identification $p_n \to i\hbar\partial/\partial x^n$ when acting on complex wave functions from which we see that field disturbances $\Phi(x)$ must at least obey a relativistic Klein–Gordon (KG) equation:

$$(p^2 - m^2 c^2)\Phi = -(\hbar^2 \partial^2 + m^2 c^2)\Phi(x) = 0. \qquad (4.11)$$

Therefore a distribution over various wave-numbers[4] as below,

$$\Phi(x) = \int \varphi(k) \exp(-ik \cdot x)\, d^4k/(2\pi)^4; \quad p \equiv \hbar k, \qquad (4.12)$$

will obey the KG equation provided $\varphi(k)$ is constrained to obey $(\hbar k)^2 = m^2 c^2$.

Before going any further it will pay us to simplify the notation and make for easier reading later on. Two fundamental natural units stand out: they are the speed of light c in vacuum and Planck's constant $h \equiv 2\pi\hbar$; these are generally believed to be invariant over all space and time so we will take them as the units of 'speed' and of 'action' respectively. Dimensional analysis tells us that all quantities in nature can be expressed as powers of mass [M], length [L], time [T] and Boltzmann temperature [K]. Because speed like c is given as $[L][T]^{-1}$ and action (momentum × position) like h is interpreted as $[M][L]^2[T]^{-1}$, we see that we can reduce all quantities to length or

---

[4]The convention here is that $d^4k \equiv dk^0 dk^1 dk^2 dk^3$ and similarly for d^4x and d^4p.

mass and Boltzmann units by taking $c = 1, \hbar = 1$ as our fundamental scales. Including gravitation one is tempted to take the Newtonian gravity constant $G_N = 1$ as another fundamental invariant and that would dispense with the remaining mass dimension, leaving only [K]. That description corresponds to using 'Planckian' units; we shall not go so far as that in this book and we will be satisfied for all quantities to be accorded a length [L] or $[M]^{-1}$ dimension in *natural units*. As an aside, we should notice that the arguments of functions like exp, trigonometric, etc., are all dimensionless.

Using natural units we can now compact the notation even more by defining a Lorentz invariant integration measure over the timelike light cone and its corresponding delta function:

$$S_{\mathbf{p}} \equiv \int d^4 p \, \theta(p_0) \, \delta(p^2 - m^2)/(2\pi)^3 = \int \frac{d^3 \mathbf{p}}{(2\pi)^3 \sqrt{\mathbf{p}^2 + m^2}}$$

(4.13)

$$\bar{\delta}^3 \, (\mathbf{p}, \mathbf{p}') \equiv (2\pi)^3 \, 2\sqrt{\mathbf{p}^2 + m^2} \, \delta^3(\mathbf{p} - \mathbf{p}') = (2\pi)^3 \, 2E_p \, \delta^3(\mathbf{p} - \mathbf{p}').$$

(4.14)

In that way we may represent a complex scalar field and its conjugate by the Fourier integrals ($p = \hbar k$):

$$\Phi(x) = S_{\mathbf{k}} \left[a(\mathbf{k}) \exp(-ik \cdot x) + b^\dagger(\mathbf{k}) \exp(ik \cdot x) \right] \qquad (4.15)$$

$$\Phi^\dagger(x) = S_{\mathbf{k}} \left[a^\dagger(\mathbf{k}) \exp(ik \cdot x) + b(\mathbf{k}) \exp(-ik \cdot x) \right]. \qquad (4.16)$$

Now comes the key point....

We *second quantise* the field by interpreting the mode $a(\mathbf{k})$ as the annihilation operator for a particle of momentum k and $b^\dagger(\mathbf{k})$ as the creation operator for an antiparticle of momentum k. (The latter can be constructed as the annihilation operator for a particle moving backwards in time, rather than an antiparticle moving forwards in time, as Feynman shows.) Taking our cue from the harmonic oscillator, we then invoke the following commutation relations for the scalar field:

$$[a(\mathbf{k}), a^\dagger(\mathbf{k}')] = \bar{\delta}^3 \, (\mathbf{k}, \mathbf{k}'), \, [a(\mathbf{k}), a(\mathbf{k}')] = 0, \, [a^\dagger(\mathbf{k}), a^\dagger(\mathbf{k}')] = 0,$$

(4.17)

$$[b(\mathbf{k}), b^\dagger(\mathbf{k}')] = \bar{\delta}^3(\mathbf{k}, \mathbf{k}'), \ [b(\mathbf{k}), b(\mathbf{k}')] = 0, \ [b^\dagger(\mathbf{k}), b^\dagger(\mathbf{k}')] = 0,$$

$$(4.18)$$

$$[b(\mathbf{k}), a(\mathbf{k}')] = [b(\mathbf{k}), a^\dagger(\mathbf{k}')] = [b^\dagger(\mathbf{k}), a(\mathbf{k}')] = [b^\dagger(\mathbf{k}), a^\dagger(\mathbf{k}')] = 0.$$

$$(4.19)$$

If the field happens to be real, then $\Phi = \Phi^\dagger$ and $a = b$, disregarding Eq. (4.19); an example of this case is the π^0 meson field. Otherwise the conjugates are deemed to carry opposite charges, like the $\pi^\pm$ fields. Again, taking our cue from the harmonic oscillator, a multiparticle state is achieved by acting with creation operators on the ground state, which is identified as the vacuum state:

$$|p_1, p_2, \ldots, p_n\rangle \equiv a^\dagger(p_1) a^\dagger(p_2) \ldots a^\dagger(p_n) |0\rangle.$$

It is automatically symmetric in its arguments for a scalar field, in view of Eq. (4.17).

4.4 Commutation relations and quantum statistics

This leads us to consider other fields besides scalar like that of the electron or photon and we will need to retrace some history. Soon after the Rutherford–Bohr model of the atom was established, with electrons occupying nuclear-bound states, it became apparent that electrons were forced to occupy successively higher energy levels for some reason. Pauli concluded that electrons with identical characteristics cannot occupy the same level and he called that the 'exclusion principle'. Because two electrons do occupy the same energy level as in the helium ground state Pauli deduced that they had to carry a two-valued property; this became recognised as spin $1/2$ (up and down) after Uhlenbeck and Goudsmit proposed it to account for the Zeeman effect in spectroscopy, amongst other evidence.

A simple explanation for the exclusion principle was proposed by Fermi and Dirac. They proposed that electrons (and likewise other spin $1/2$ particles like nucleons) obey anticommutation relations, instead of commutators, for operators like $a, a^\dagger$. Thus for spin $1/2$ objects they stated that in place of (4.17) the field

Fig. 4.8   Pieter Zeeman.   Fig. 4.9   Paul Adrien Maurice Dirac.

operators obey

$$\{a(\mathbf{k}), a^\dagger(\mathbf{k}')\} = \bar{\delta}^3(\mathbf{k}, \mathbf{k}'), \quad \{a(\mathbf{k}), a(\mathbf{k}')\} = 0, \quad \{a^\dagger(\mathbf{k}), a^\dagger(\mathbf{k}')\} = 0 \tag{4.20}$$

where the anticommutator is defined to be $\{A, B\} \equiv AB + BA$. This means that creation of a two-particle state like

$$|p_{1\Downarrow}, p_{2\Uparrow}\rangle \equiv a_\Downarrow^\dagger(p_1) a_\Uparrow^\dagger(p_2)|0\rangle$$

is automatically antisymmetric in its arguments. In particular if two electrons are in the ground state momentum configuration, they have to be in a spin antisymmetric state, $|\Uparrow\Downarrow\rangle - |\Downarrow\Uparrow\rangle$, making for zero total spin. It also means that one cannot fit an extra electron into the ground state as the anticommutation relations lead to zero since the square $(a^\dagger(p))^2 = 0$ whether the spin is up or down. The Pauli exclusion principle is guaranteed.

Fig. 4.10 George Uhlenbeck.

Fig. 4.11 Samuel Abraham Goud-smit.

A proper proof that commutators apply to integer spin particles, whereas anticommutators apply to half-integer spin particles, is too technical for this book. This is referred to as the 'spin-statistics' theorem and was first mooted by Fierz and Pauli [11]; We call integer spin objects *bosons* and half-integer spin objects *fermions* to accord with names of the originators of the statistics which they obey. The repercussions of the commutation rules for counting multiparticle states are quite striking too and lead us into pastures new: the subject of quantum statistics, which lies outside the scope of this monograph but can be consulted in the physics literature. Basically, what happens is that for a system with energy E the Bose–Einstein distribution mentioned in Eq. (4.2), namely $1/[\exp(E/\mathfrak{k}T) - 1]$ gets replaced by the Fermi–Dirac distribution $1/[\exp[(E - F)/\mathfrak{k}T + 1]$. F is called the 'Fermi energy' and it is the maximum energy of fermions (at zero temperature) contained in an enclosure as fermion states get populated from the ground up. It is true to say that *bosons like to be in the same state and cooperate* (which can lead to Bose–Einstein condensation at low temperatures) whereas *fermions like to keep their distance and stay apart* (and thereby lead to atomic stability). Perhaps these remarks about particle gregariousness mimic the behaviours of political party adherents!

Chapter 5

The Grand Communicators

Before going any further we need to appreciate a few things. Communication is the transmission of information or exchange of property between one system and another. In quantum field theory this is achieved by the exchange of at least one force field quantum. Such disturbance quanta $\Phi(x)$ in and of the medium (even the vacuum) obey field equations of one sort or another. When they satisfy the wave equation $\Box\Phi = 0$ they lead to transmission with the speed of light and correspond to particles of zero mass; the communication is as fast as possible but is *not* at infinite speed. In turn it leads to a 'potential' $V(r) \propto 1/r$ between two systems separated by r which is 'long range' as in electricity: it continues to be felt at large r even though it has been weakened greatly with distance. See Section A.2 for the explanation.

On the other hand, when the quantum of transmission obeys the massive KG equation $(\Box + m^2)\Phi(x) = 0$, corresponding to a communicator of mass m, it leads to a Yukawa potential $V(r) \propto \exp(-mr)/r$ which fades extremely rapidly with distance from a point source placed at the origin (see Section A.2 again). It is said to be 'short range' since it is miniscule at distances much larger than about $1/m$. This is what happens with the transmission of the so-called weak boson force, so much so that Fermi thought weak interactions were due to contact interactions. The surprising thing is that in strong QCD the result (from quantum effects) is that the colour force *increases* more or less linearly with distance and prevents separation of the two systems which are therefore tied in an inescapable

embrace through the strong gluons. We shall explore this feature
later on.

We now need to ask how the dynamics leads to these particle
exchanges, carrying the communication. Classical mechanics has a lot
to teach us in this regard. Its most glorious manifestation is through
the Lagrangian or Hamiltonian formalisms. The Lagrangian L carries all the dynamics of a system. It is a function of the essential
dynamical variables q or 'coordinates' that characterise a particular
system. L is basically the difference between the kinetic and potential energy, expressed in terms of q. Normally the 'kinetic' part is
quadratic in the time derivative or associated 'velocity' $\dot{q}$, while the
potential energy $V(q)$ specifies how the 'static' part of the energy
varies with q. Once $L(q, \dot{q})$ is worked out or stated, the dynamics of
the entire system is fully determined.

The dynamical Euler–Lagrange equations of motion — the generalised versions of Newton's equation for force driving acceleration —
are arrived at most elegantly by constructing the so-called 'action'
which is the integral $\int L(q, \dot{q})\, dt$ over some time interval and minimis-
ing it by Hamilton's principle; the resulting Euler–Lagrange equation
encapsulating the dynamics produces second order in time derivatives
of the variable q[1]:

$$\frac{\partial}{\partial t}\left(\frac{\partial L}{\partial \dot{q}}\right) - \frac{\partial L}{\partial q} = 0 \qquad (5.1)$$

How does this translate into field theory? Well, the Lagrangian
has to depend on the field variables Φ which pervade space, so L is
an integral over space of a 'Lagrangian density' and we can infer that

$$L = \int \mathcal{L}(\Phi(x), \partial\Phi(x))\, d^3x$$

with the corresponding field equations:

$$\partial_n\left(\frac{\partial \mathcal{L}}{\partial_n \Phi}\right) - \frac{\partial \mathcal{L}}{\partial \Phi} = 0. \qquad (5.2)$$

---

[1] By contrast, the Hamiltonian formalism which uses q and its corresponding
momentum $p \sim \dot{q}$ leads to two coupled first-order equations involving $\dot{q}, \dot{p}$.

One can also conclude that for a free field of mass m the choice,

$$\mathcal{L} = [(\partial^n \Phi)(\partial_n \Phi) - m^2 \Phi^2]/2$$

will yield the massive Lorentz invariant KG equation: $(\Box + m^2)\Phi = 0$.
We can regard the second term of the equation above as the mass
contribution of potential energy and the first term as the contribution
of kinetic energy of the field since it involves the square of the field
derivative and is associated with the (momentum)^2, viz. $p^2\, a^\dagger(p)a(p)$,
carried by the creation and annihilation operators.

In a sense all of the above is fairly insignificant, since *nothing
happens*: the free fields amble along without any mutual interactions
and the states remain the same for all time. The system is devoid
of events! Scientific interest is awakened only after we include the
disturbances to the various fields, which in some ways impact on the
nature of time itself. In Section A.3 we have given an example of a
cubic interaction between two independent fields and how the effects
are estimated by perturbation theory and pictured through the use
of Feynman diagrams. This is a huge subject in its own right and
would take us too far afield from the main theme of this monograph,
so I recommend readers who want to pursue this topic to consult one
of the many excellent textbooks on quantum fields and particles, for
instance the texts by Weinberg [12].

5.1 Gravitational wobbles

Empty space is believed to be flat and, as it is devoid of mass/energy,
it is described by a Lorentz invariant metric η_{mn}. As Einstein taught
us, introducing masses will distort spacetime and if the masses interact with one another through the gravitational force we may expect
the spacetime to be disturbed accordingly. If the distortion is sufficiently severe we should expect gravity waves to be generated; this
is in analogy to waves in water caused by a tsunami say.

The manner by which this is implemented is to write the full
metric characterising disturbed spacetime as $g_{mn} = \eta_{mn} + \kappa h_{mn}$ and
see what happens when one invokes the Einstein–Hilbert pure gravity

Fig. 5.1   David Hilbert.   Fig. 5.2   Marcel Grossmann.

Lagrangian,

$$\mathcal{L} = 2R^{[g]}\sqrt{\det(-g)}/\kappa^2. \tag{5.3}$$

$R^{[g]}$ is the Ricci spacetime curvature scalar which needs to be minimised by the 'action principle' (4.2) with h playing the role of Φ; $\kappa^2 \equiv 32\pi G_N$ and G_N is the Newtonian gravity constant. I am not going to explain how $R^{[g]}$ (which is determined by the full metric g) produces an expansion of $\mathcal{L}$ in powers of κ, but will refer you to textbooks on general relativity [13, 14] as the argument and mathematics are quite complicated. The important point to grasp is that the metric becomes a field in general relativity and the quantised disturbance, like h_{mn}, becomes the quantum 'graviton'. [Maths-shy readers can skip the next paragraph.]

The upshot of the exercise is that we get an infinite series of terms from (5.3). Fierz and Pauli [15] determined some of these terms. The series starts off quadratically in h with

$$\mathcal{L}_0 = (\partial^l h^m{}_m \partial_l h^n{}_n - \partial_l h^{mn} \partial^l h_{mn})/2 + \partial_l h^l{}_m \partial^n h_n{}^m - \partial^m h_{mn} \partial^n h^l{}_l, \tag{5.4}$$

and this is sometimes taken as the starting point for the kinetic energy of a model of massive gravity. The higher order in κh terms (cubic, quartic, etc.) represent the mutual interactions of the gravitational disturbances, and they make perturbation analysis messy and difficult. Be that as it may, $\mathcal{L}_0$ is unaffected by the 'gauge transformation' $h_{mn} \to \partial_m \theta_n + \partial_n \theta_m$ which corresponds to 'general covariance' under coordinate changes. Of course h, like g, is symmetric in its indices and now represents spin 2: the disturbances are tensorial or of *quadrupole* type, *not dipole* like magnetic nor *monopole* like electric.

Quantising the disturbance h_{mn} produces the graviton as the quantum of gravitational communication. Although it is highly improbable that a single graviton will ever be detected because its interaction is so very weak, we may still envisage the gravitational force between two masses as the continual exchange of gravitons between them. What is especially interesting is the speed of communication, derivable from (5.4). The absence of mass terms indicate that it is at the speed of light, and that the graviton is massless like the photon. It was a great comfort to theorists when the gravitational wave signal from two merging neutron stars arrived at the same time as the gamma ray flash arising from the coalescence, despite the vast distances involved. It meant that massive gravitons, which some theories propose, are excluded.

5.2 Electromagnetic propagation

Unlike gravitons, the quanta of electromagnetism, namely the photons, can sometimes be captured singly. To see photons individually requires electronic aids: a photomultiplier tube can be used to enhance the signal through the photoelectric effect. (It is sometimes claimed that a dark adapted human eye can spot one photon, but this has not been convincingly proven.) The construction of single photon sources is highly sophisticated, requiring solid state devices (like CCDs) or through damping of laser beams. At the high energy end of the spectrum, γ rays are detected through scintillators which convert the very high frequencies to visible ones.

The 4D electromagnetic potential A_m is a real field and is the primary quantity which is being quantised, with repercussions for the Maxwell tensor $F_{mn} \equiv \partial_m A_n - \partial_n A_m$. It might seem that the quanta of A, being vectorial, possess four physical degrees of freedom, but this conclusion is incorrect. In fact photons possess two degrees of freedom or polarisations, not four. Because of the gauge symmetry under $A_m \to A_m + \partial_m \theta$, any component of $A(k)$ that is parallel to the momentum k is totally unphysical; lastly because any current $J_m = \partial^n F_{nm}$ to which A couples is automatically conserved ($\partial^m J_m = 0$), the longitudinal component $\mathbf{k}$ drops out too. This leaves us with two polarisation components $A_\Uparrow, A_\Rightarrow$ orthogonal to the direction of propagation of the wave, which remain physical. Another way of saying this is that the only physical degrees of freedom are the right helicity and left helicity states having magnitude of spin $\hbar$. Basically the other two longitudinal components cancel out between one another in interactions.

Actually a similar argument applies to the previously mentioned graviton field h_{mn} which for a symmetric field seems to have 10 degrees of freedom at first sight. However the freedom to choose a particular coordinate system (which corresponds to general covariance of the equations of motion) removes four degrees of freedom, whilst another four disappear because of conservation of the associated current, which is nothing but the energy–momentum stress tensor. So effectively h_{mn} has only two physical degrees of freedom and these correspond to states of right and left helicity, with spin magnitude $2\hbar$.

It is quite astonishing how much humans have exploited the electromagnetic spectrum for their benefit; from radio waves, through to microwaves, infrared, ultraviolet, X-ray and gamma ray frequencies. The communication takes place at light speed, which is effectively instantaneous so far as human experience is concerned. Atoms rely on photon exchange to produce the inverse radius potential law that binds electrons to the nuclear core. Atoms get their stability through the Pauli principle, arranging electronic orbits in the form of 'shells'. The shell occupation numbers distinguish one element from another. Each photon frequency energy range helps us to understand the workings of nature and discover new information. Radio waves are ubiquitous for human communication, shaping our society in ways undreamed of. Microwaves and lasers are to be found in every home in daily use, while X-rays have greatly influenced medical developments. There are many other examples of electromagnetic wave incursion into our daily lives and it is quite amazing to think that none of these changes could have been contemplated 200 years or so ago.

5.3 Nuclear messaging

Round about the mid-twentieth century it was understood that the nuclear force between two nucleons was primarily due to the exchange of a massive *pseudoscalar* pion field (mass μ say); it is short range and changes sign under space reflection. This leads to a strong potential $V(r) = \exp(-\mu r)/4\pi r$, as proved in Section A.2. It is associated with an interaction $\mathcal{L}_g \sim g\bar{N}\pi N$ with a large coupling constant $g^2/4\pi \sim 15$ in contrast with electromagnetism strength where $e^2/4\pi \simeq 1/137$. That picture fits in very well with our description of nuclear physics whereby nucleons come in two varieties $N = (p, n)$, forming a *isodoublet* representation of a fictitious 'isospin' group. The nucleons interact with an *isotriplet* representation of pions, (π^+, π^0, π^-). It was thought at the time that the only distinction between the proton and neutron lay in their different electromagnetic characteristics, so if the latter were ignored the properties of p and n would be the same. The use of the isospin group which 'rotates' the isospin multiplets is thought to be the approximate symmetry of nuclear physics and it does serve to classify the nuclei nicely.

Pions are not stable but decay rather quickly. Charged pions can make tracks in emulsions and have lifetimes of the order of 10^{-8} seconds. The charged pions will decay primarily into muons, which are relatively long-lived, at high speed; but the π^0 decays very much faster into a pair of gamma rays that share the mass of the pion.

With the advent of fundamental quarks, this picture has now been revised drastically. The proton is now considered to be a bound state of three quarks coming in a white colour combination of type

$(u_{\text{red}} u_{\text{green}} d_{\text{blue}})$ while the neutron is regarded as $(u_{\text{red}} d_{\text{green}} d_{\text{blue}})$.
Both combinations are bound in a soup of colour 'gluons'. It turns
out that the mass difference between p and n is not only caused by
their different charges but by an intrinsic mass difference of about
$m_d - m_u \simeq 4$ MeV between trapped u and d quarks. The colour
force governing the interaction between the quarks gets stronger as
the quarks get separated by r; it is roughly given by a linear and
Coulomb potential: $V(r) \simeq \alpha/r + Kr$ and there is so much stored
vacuum energy between the quarks that at a critical separation the
vacuum decomposes into a quark–antiquark pair. Figures 5.3 and 5.4

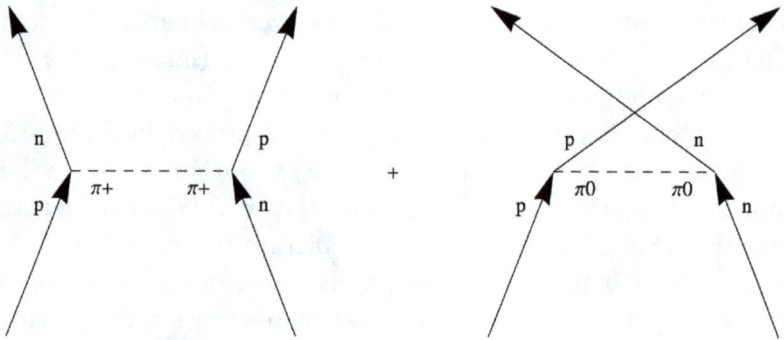

Fig. 5.3   Proton–neutron binding through pion exchange.

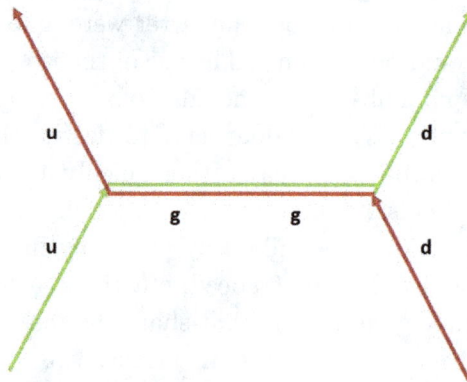

Fig. 5.4   Quark binding u–d through gluon g exchange.

contrast the old isospin picture and the new colour picture of the strong force. Nonetheless the SU(2) isospin group, largely promoted by Fermi and Wigner, remains a very useful tool for studying nuclear states and nuclear interactions at higher level.

5.4 Weak signals

As mentioned in Chapter 2, the weak interactions are communicated by very short range forces associated with massive $W^{\pm}$-bosons and a Z^0-boson. These three bosons are pretty heavy ($\sim$80 to 90 GeV) and have extremely short lifetimes of the order of 10^{-25} seconds, which is inferred by their rapid decays into leptons. They are the particles responsible for beta decays and for rare neutrino elastic scattering.

The weak force is connected with an SU(2) gauge group again, acting on lepton doublets like (ν^0, e^-) and quark doublets like (u, d). Normally such a group would lead to a triplet of *massless* gauge bosons, which the weak bosons are clearly not! The mystery as to how and why W and Z acquire mass is resolved by the process

Fig. 5.5   Enrico Fermi.

Fig. 5.6   Eugene Wigner.

of 'spontaneous symmetry breakdown': a scalar boson (the Higgs) acquires a vacuum expectation value, and its coupling to the gauge bosons allocates them their masses. This is a fascinating tale in its own right and its full explanation is the stuff of legend requiring separate story-telling. The tale ended with the discovery in 2015 of the Higgs–Brout–Englert quantum accompanying the scalar background field — the Higgs boson with mass of about 125 GeV. Section A.4 sketches a simpler model than the Standard Model, just based on the gauge group $U(1)$; it contains the seeds of the idea and the mathematically equipped reader is invited to go over the details.

In all of these cases we see that the grand communicators are gauge field exchanges, vectorial or tensorial. Gauge symmetry seems to be an overarching principle governing the forces of nature: that the Lagrangian takes the same form no matter how we phase the fields into one another from place to place.

Chapter 6

Get the Picture

The last five chapters, augmented by the Appendix, have been in the nature of a refresher course on quantum fields for those of you who have come across that topic. And for those of you unfamiliar with the subject, those chapters have tried to convey the basic ideas about particles and their associated fields. It is now time to introduce concepts which underlie the entire motivation for this monograph and how they manage to produce a unified picture of all the known natural forces.

6.1 Anticommuting coordinates

There are two types of fields in physics: those connected with integer spin particles, called bosons, and those connected with half-integer spin particles called fermions. The former fields Φ have the property of commuting with one another when their arguments are space-like separated (so that they cannot impinge on one another even through light), while the latter Ψ anticommute under those circumstances. Thus for $(x - x')^2 < 0$,

$$[\Phi(x), \Phi(x')] = 0, \quad \{\Psi(x), \Psi(x')\} = 0. \tag{6.1}$$

The spin-statistics theorem [11] provides the connection between Eq. (6.1) and physical fields in question. Of course in the interpretation of that connection, the locations x, x' are ordinary commuting parameters.[1]

---

[1] As they refer to spacetime. There have been proposals to quantise the locations themselves since at extremely short distances precise positions are unrealisable by

To break the shackles of the spin-statistics connection means that the unconventional fields are unphysical and must not appear in the physical state spectrum: they cannot exist in the initial and final states, but are allowed to appear in intermediate configurations. That picture is exploited when quantising gauge fields by the Faddeev–Popov–BRST method [16]: one introduces fields which defy the spin-statistics theorem but only occur mid-calculation, disappearing at the beginning and end. They are called 'ghost fields' and are necessary to ensure physical probability conservation; they serve to cancel out unphysical degrees of freedom in the gauge field communication. Lorentz scalar ghosts therefore obey anticommutation relations and are accompanied by *anticommuting parameters*.

The concept of anticommuting variables originated with Grassmann's development of geometry and exterior algebras. He envisaged a set of variables $\xi, \eta, \zeta, \ldots$ that obey a series of anticommutation rules:

$$\{\xi, \xi\} = 0, \text{ so } \xi^2 = 0, \quad \{\eta, \eta\} = 0, \text{ so } \eta^2 = 0,$$
$$\{\xi, \eta\} \equiv \xi\eta + \eta\xi = 0, \ldots$$

These variables are effectively fermionic. They commute with bosonic coordinates like x; we are allowed to add and multiply Grassmann variables provided that any relationship which we care to write is uniformly bosonic or fermionic with each term, *not a schizophrenic mixture of terms*.

Let me now describe some of their properties; they are really quite simple and the reader shouldn't have any difficulty in following the elementary algebra they enjoy. Assuming they are all real so $\xi^\dagger = \xi$, etc., we should point out that the product $\xi\eta$ is imaginary because its hermitian conjugate[2] $(\xi\eta)^\dagger = \eta^\dagger\xi^\dagger = \eta\xi = -\xi\eta$. Thus $i\xi\eta$ is real and bosonic since the product commutes with the other variables like ζ. An expansion of a quantity as a polynomial in such variables has

---

the uncertainty principle, so spacetime changes into some sort of foam. However nothing concrete has come out of such an idea apart from the suggestion that quantum field infinities may disappear in that regime.

[2]We denote hermitian conjugation by † for which there exists the conjugation rule, as for matrices, which reads $(AB)^\dagger = B^\dagger A^\dagger$.

to *terminate*, ending with the maximum power $\xi\eta\zeta\ldots$ of as many variables as there are. From now on I shall use late letters of the Greek alphabet to describe them and reserve Latin letters for bosonic coordinates. I must also point out that any product of anticommuting variables is nilpotent, meaning that its square vanishes identically, e.g. $(\xi\eta)^2 = \xi\eta\xi\eta = -\eta\xi\xi\eta = 0$.

We may also complexify combinations of these ξ, in analogy with our standard treatment of commuting coordinates; thus the complex variable construction for bosonic coordinates x, y, z,

$$z = x + iy, \quad \bar{z} = x - iy,$$

has as its Grassmannian counterpart:

$$\zeta = \xi + i\eta, \quad \bar{\zeta} = \xi - i\eta,$$

leading to $\bar{z}z = x^2 + y^2$ but $\bar{\zeta}\zeta = 2i\xi\eta$ instead. As we shall be coming across a number N of independent complex ζ, I shall enumerate them as ζ^μ where μ runs from 1 to N. The conjugates will be labelled $\zeta^{\bar{\mu}}$ rather than $\bar{\zeta}_\mu$, as is customary, in order to adhere to Einstein notation; this leads one to the phase invariant sum

$$Z \equiv \zeta^{\bar{\mu}}\zeta^\mu = \zeta^{\bar{1}}\zeta^1 + \zeta^{\bar{2}}\zeta^2 + \cdots + \zeta^{\bar{N}}\zeta^N$$

which is unchanged under the full unitary group $U(N)$ actions on the N ζ-terms in fact.[3]

Now comes the intriguing part, concerning the calculus of these anticommuting variables. As is conventional I shall take all derivatives as acting from the left in order not to confuse readers with right differentiation and its complications. The differentiation operator $\partial/\partial\zeta$ is also fermionic, so we must pay careful attention to the order of the terms on which it acts. Thus

$$(\partial/\partial\xi)\chi\xi\eta = -\chi\eta = \eta\chi, \text{ etc.}$$

Very curious is how Grassmann integration works. The rules for integrating over anticommuting variables were established by

---

[3] Actually Z admits a $2N$ symplectic group because for each ζ, we encounter the skew product $\bar{\zeta}\zeta = I(\xi\eta - \eta\xi)$.

DeWitt [17] and Berezin [18] and discussed systematically by Rogers [19]. They proved that for consistency we must take $\int d\zeta \cdot 1 = 0$, $\int d\zeta \cdot \zeta = 1$, so in some sense they correspond to taking the derivative and then setting the variable to zero. It is quite bizarre in as much as an integral such as $\int d\zeta \, (\partial/\partial\zeta) \, \zeta\chi = 0$ means that the 'endpoints' of the integral contribute nothing. These rules for differentiation and integration of anticommuting parameters take some getting used to but they are completely consistent as it happens.

In later chapters we shall be visiting integrals over N complex variables and their conjugates: $\zeta^1\zeta^2 \cdots \zeta^N$, $\zeta^{\bar{N}} \cdots \zeta^{\bar{2}}\zeta^{\bar{1}}$. To get a nonzero answer from the integration for the full integral, we need the complete product of all the variables. For example consider a polynomial over just a couple of variables. Let $\mathcal{S}$ be a symmetric matrix whose elements are bosonic, for instance

$$\mathcal{S} = \begin{pmatrix} a & b \\ b & c \end{pmatrix}.$$

Then the integral over a two-component coordinate (x, y) over the whole plane gives

$$\iint dx dy \exp\left[-(x, y)\, \mathcal{S} \begin{pmatrix} x \\ y \end{pmatrix}\right] \propto [\det(\mathcal{S})]^{-1/2}, \qquad (6.2)$$

whereas with a simple antisymmetric matrix $\mathcal{A}$ acting between a two-component fermionic doublet;

$$\mathcal{A} = \begin{pmatrix} 0 & -d \\ d & 0 \end{pmatrix}$$

leads one to the integral

$$\iint d\xi d\eta \cdot \exp\left[-(\xi, \eta)\, \mathcal{A} \begin{pmatrix} \xi \\ \eta \end{pmatrix}\right] = 2d \propto [\det(\mathcal{A})]^{1/2}, \qquad (6.3)$$

Perceptive readers will notice the change of sign in the exponents of the determinant between Eqs. (6.2) and (6.3). This is a general feature when we compare commuting and anticommuting coordinates as we will do in the next section.

With N anticommuting complex ζ and the summed $Z \equiv \zeta^{\bar{\mu}}\zeta^{\mu}$ the integration over all ζ requires us to take Z^N in order to get a nonzero result; any other power of Z gives zero. Thus given the DeWitt–Berezin rules, it is a simple exercise to show that the N-fold anti-integral,

$$\int (d\zeta^1 d\zeta^{\bar{1}})(d\zeta^2 d\zeta^{\bar{2}}) \cdots (d\zeta^N d\zeta^{\bar{N}}) \, Z^M = N! \, \delta^{MN}, \qquad (6.4)$$

a conclusion which we will make ample use of later.

6.2 Dimensions, positive and negative

We are all familiar with the notion that we live in a space of three dimensions, meaning that we are free to roam independently over three degrees of freedom: right–left, forward–backward, up–down. We assign to them three coordinates: x, y, z in the traditional Cartesian sense, though in depicting them on paper we are limited to two degrees of freedom by flattening out our perspective. However if we give our imagination free rein we can conceive of other new directions by adding further independent directions, such as time, to give us the full 4D spacetime. Every time we add a bosonic coordinate we open a door into a new vista with all its possibilities. However if we have look at Table 6.1 — where we contrast Bose–Einstein variables with Fermi–Dirac variables — the converse is true: for each

Table 6.1 Comparison between Bose–Einstein and Fermi–Dirac variables.

Characteristic	Bose–Einstein	Fermi–Dirac
c-number vs a-number	$xy = +yx$	$\xi\eta = -\eta\xi$
Commutation relations	$[a, a^\dagger] = 1$	$\{a, a^\dagger\} = 1$
At spacelike $x - x'$ separation	$[\Phi(x), \Phi(x')] = 0$	$\{\Psi(x), \Psi(x')\} = 0$
Statistical distribution	$(e^{\alpha+\beta E} - 1)^{-1}$	$(e^{\alpha+\beta E} + 1)^{-1}$
Negative dimensional integrals	$\int d^{-N}x$	$\int d^N\xi$
Group representations	O$(2n)$ v Sp$(-2n)$	Sp$(2n)$ v O$(-2n)$
Quadratically *infty* loop integrals	Bose loops	Fermi loops

anticommuting coordinate that we introduce we close a door! It is
not too far-fetched to say that with each BE coordinate we add a
dimension, but with each FD coordinate we subtract a dimension;
put another way, bosonic dimensions act positively while fermionic
ones act negatively.

To reinforce this argument, Halliday and Ricotta [20] noticed that
if one continued the dimension of the bosonic integration variables
from positive to negative, then the result matched the use of positive fermionic integration. Group theoretic considerations promote
this view even further. King [21] observed that the dimensions of
irreducible representations of $\mathrm{Sp}(2N)$ (the Lie group characterising
fermionic variables) were the same as those of $\mathrm{SO}(2N)$ (the group
characterising bosonic variables) with $N \to -N$ and the Young diagrams transposed. Even the sizes of those groups in question, namely
$N(2N-1)$ and $N(2N+1)$ share that property. Last but not least the
infinite Feynman integrals over bosonic loops can be counterbalanced
by those from fermionic loops because the latter come with a nega-
tive sign, due to tracing over a squared bilinear such as $(\bar{\Psi}\Psi)(\bar{\Psi}\Psi)$.
No doubt experts can point to other examples when fermionic coor-
dinates diminish the nett dimensions that have been added to by
bosonic coordinates. As Krauss [22] has noted in his celebrated book
"A Universe from Nothing", nature abhors a vacuum: *nothing* is
unstable. It is even conceivable that the universe started off with
zero dimensions and budded into bosonic and fermionic spaces with
cancelling dimensions, leaving zero overall; this resembles the claim
that the total energy of the universe is zero, with positive kinetic
energy counterbalanced by negative potential energy (though this
may be controversial).

6.3   Where–when–what events

I now come to the central theme of the book, the need to describe
events *fully*. This is where I try to break the shackles of traditional
descriptions. The first thing to realise is that the natural state of
the universe is one of commotion: its expansion of course and its
evolution from a state of hot turmoil into one of cold uniformity. A
static configuration is inconceivable since it would mean the cessation

of all events, so we would be unaware of it; communication would end and even the notion of time itself would become problematic. We should rejoice that the passage of time is punctuated by the occurrence of events, which set up a frequency standard when we are convinced of their regularity.

The second point to note is that an event at a given location and time is incomplete without specifying the nature of the change and the participants at the event. We are very familiar with Cartesian coordinates that say where an event has occurred and we even report when the event took place. But as to what happened... this depends on its nature. After we are wise as to what evolved, in quantum field theory we combine the participating fields (which often belong to a representation of some group) into an interaction that is local and is Lorentz invariant in order to satisfy the tenets of special relativity. It seems to me that what we require is a set of coordinates to specify the properties of the interacting fields. Such properties are called quantum numbers and correspond to eigenvalues of some 'internal group' so those property coordinates have to carry those quantum numbers. Since experiments suggest that the number of quantum attributes is finite, so must the property coordinates be.

To make analogies with ordinary life, here are four examples:

- Suppose we are storing a file on a computer. A decent operating system will specify the file size (its space), the timestamp (its time) and its type (its attribute or extension) so that others can later decipher and display it. These days there are only a finite number of these file attributes (quantum numbers).
- Suppose a reporter on the news is describing a murder. The murder is at some site (space) and has occurred within some (time) interval and is of some type (nature of death or attribute). Presumably, there are only a finite number of ways in which death could have been administered and might conceivably be ascribed to property coordinates.
- A military historian annotating battles would specify their location on Earth (space) and the date on which they took place (time) and the combatant nationalities (attributes).

- A geographer describing catastrophic natural events would specify their location on Earth (e.g. Sunda Strait), the date of the
  catastrophe (e.g. August 1883) and the type of catastrophe (e.g.
  volcanic eruption of Krakatoa).

   While it is obvious how to specify a location and instant of an
event through the use of spacetime coordinates, it is far less obvious
how to specify attributes mathematically. The way this idea came
to me was only after we showed that gravity and electromagnetism
could be unified very simply by introducing a single Lorentz scalar
anticommuting variable ζ associated with charge — the property of
carrying charge signifying 'electricity'. Once that was understood,
it was a matter of enlarging the concept to other quantum number
carrying ζ. In particular the quantum numbers associated with neutrinos can be dubbed 'neutrinicity'; likewise the three ζ (red, green,
blue) needed to describe QCD colour can be termed 'chromicity'.
   Why make the liaison between (complex scalar) property with
anticommutation? Because the squares of each ζ vanish, so once a
system possesses one such attribute, it cannot doubly have it. For
instance a surface can be red-reflecting (meaning it absorbs other
colours) and that's it — it cannot be doubly ζ_{red}. The choice of
complexity is because a surface can be the conjugate, $\bar{\zeta}_{red}$, namely
red-absorbing (so it looks cyan if it reflects the other colours). One
may also combine the properties and their conjugates, like the prod-
uct $\bar{\zeta}_{red}\zeta_{red}$ in which case it is neither red-absorbing, or red-emitting
so it is black or 'neutral'. The question is: "Why make the ζ scalar
rather than carrying spin", which would be much more natural and
fully consistent with the spin-statistics theorem? The simple answer
is: "Because addition of spin leads to spin proliferation", as I shall
be discussing at greater length when we later come to conventional
supersymmetry.[4] For our purpose, scalar ζ coordinates are simply
aids for describing attributes and escaping spin-statistics strictures.

---

[4]The problem of spin proliferation was pointed out to me by Gell-Mann and
becomes much more severe when one considers N-extended supersymmetry.

In order to make the idea more tangible and more amenable to illustration, in Chapter 10 I will picture the attributes as connected to human characteristics and attach them to smileys; this will make the idea seem more personal and is more fun to analyse. Imagine the following opposite pairs of human traits: Happy/Sad, Pleased/Angry, Optimistic/Pessimistic, Rich/Poor,... We could represent these by the property coordinates: $\zeta_{happy}/\bar{\zeta}_{happy}$, etc., since conjugates (or barred coordinates) correspond to the opposite attributes. Somebody could be a mixture of these characteristics: *Happy* and *Rich* but *Pessimistic* still and temporarily *Angry* — thereby painting the personality in question at the time. We might ascribe the mixed property $(\bar{\zeta}_{happy}\zeta_{rich}\bar{\zeta}_{optimistic}\bar{\zeta}_{pleased})$ to that person. However another person might be a bit of both; for example Rich but both Pessimistic and Optimistic $(\zeta_{rich}\bar{\zeta}_{optimistic}\zeta_{optimistic})$, thereby having no strong feelings one way or the other. Or to take another example: Rich and Poor (so Middle-class) but Happy $(\zeta_{happy}\bar{\zeta}_{rich}\zeta_{rich})$. I'll be using smileys to portray such mixtures later on, but the important point is that the full personality is a melange of a *finite* number of such characteristics. It is these finite melanges which solve the 'generation problem' whereby quarks and leptons get repeated.

6.4   Space–time–property diagrams

This concept can be taken over to the particle menagerie with its various quantum numbers. For instance consider the strong QCD interaction [10] depicted as a Feynman diagram in Figure 5.13, reading time from left to right, so it corresponds to the process $\bar{u}_{red}u_{green} \rightarrow$ gluon $\rightarrow \bar{d}_{green}d_{red}$. It invites us to define a quartet of property coordinates $\zeta_{red}, \zeta_{green}, \zeta_{up}, \zeta_{down}$ so that the $\bar{u}_{red}$ quark is associated with product $\bar{\zeta}_{red}\bar{\zeta}_{up}$ and similarly for the other quark states. The intermediary gluon can be identified with the combination $\zeta_{red}\bar{\zeta}_{green}(\bar{\zeta}_{up}\zeta_{up} + \bar{\zeta}_{down}\zeta_{down})$. And this already suggests the possibility of generations since a red quark can conceivably also be associated with the property combination $\zeta_{red}\bar{\zeta}_{green}\zeta_{green}$, quite aside from its flavour property (u or d). Likewise there can be two up quarks with the attributes ζ_{up} or $\zeta_{up}\bar{\zeta}_{down}\zeta_{down}$, regardless of colour.

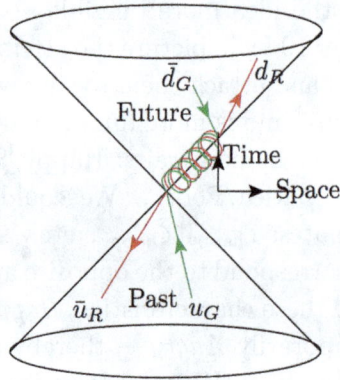

Fig. 6.1 The annihilation process $\bar{u}_{\text{red}} u_{\text{green}} \rightarrow \bar{d}_{\text{green}} d_{\text{red}}$, mediated by a gluon.

We are going to study this systematically later but is an encouraging step towards understanding the existence of particle generations.

On a related matter, how can we illustrate this process of gluon communication? Relativists have shown the way by picturing 'where–when' on a 2D projection. It is conventional to draw the time coordinate upwards and the space coordinate x sideways. A photon moving at the speed of light follows a 'light cone' if it originates at the origin since $x = ct$ with $c \rightarrow 1$ using natural units. The massless gluon does the same. However a massive object like an up quark lies within the cone and traces a line within it; likewise for a massive down quark. Thus the annihilation process mentioned above can be drawn on the same figure with coloured lines characterising their chromicities. (See Figure 6.1.) The gluon which couples to the red–green combination serves as an intermediary communicator and is drawn as a coiled double-pair (red and green). So in this case we have extended the drawing to 'where–when–what'. We can do something similar with other properties. The burning question arises: "how many properties are needed to characterise the particle zoo?" We shall address this next.

Chapter 7

Property Transformations

The critical question was raised in the last section as to how many attributes are required to describe the known elementary constituents. As always we have to be guided by experiment but must also pay heed to economy. If we invoke an abundance of different ζ then we are simply undermining the entire property rationale. Thus with the establishment of QCD as the communicator of strong force it is quite obvious that we require three distinct colours, normally taken as red, green and blue, which we will attach to ζ^1, ζ^2 and ζ^3 respectively; we cannot do with less. On the other hand, electroweak forces indicate that the leptonic pair (ν^0, e^-) are two components of a weak isodoublet, and this has nothing to do with QCD, so let us include two more ζ: namely the pair (ζ^0, ζ^4). As to what quantum numbers are carried by each of these properties, that is a separate issue which we will deal with when we come to the gauge field mediators and their couplings but we can already guess what they are. Furthermore the connection with handedness of the weak isodoublet is really important and will be covered later since it underpins the Standard Model.

7.1 Electromagnetic phases

For now we want to discuss the group theoretical aspects which emerge from these ζ bases. If there is just one property, such as electricity, we need just one ζ^4 (or simply ζ as the labelling becomes superfluous). A single complex variable $z = x + iy$, with the invariant

$\bar{z}z = x^2 + y^2 = r^2$, leads one to envisage rotations

$$\begin{pmatrix} x \\ y \end{pmatrix} \rightarrow \begin{pmatrix} x' \\ y' \end{pmatrix} = \begin{pmatrix} \cos\theta & -\sin\theta \\ \sin\theta & \cos\theta \end{pmatrix} \begin{pmatrix} x \\ y \end{pmatrix}$$

which do not alter r, or the modulus of z. This is an example of a 2D 'orthogonal' transformation between 2 real variables x and y, going by the name $\mathcal{O}(2)$. [In three dimensions it would correspond to the full 3D rotation group $\mathcal{O}(3)$.] But in the complex plane it just amounts to a change of phase, $z \rightarrow z' = \exp(i\theta)\,z$, as the reader can readily check. In this alternative view it corresponds to a one-dimensional 'unitary' transformation of a complex variable leaving $\bar{z}z$ invariant. Since the $\mathcal{O}(2)$ transformations rotate $x \rightarrow y$ and $y \rightarrow -x$, in the complex plane, the angular momentum operator

$$iJ_{[xy]} = x\frac{\partial}{\partial y} - y\frac{\partial}{\partial x}$$

performs that action and is antisymmetric under $x \leftrightarrow y$.

We can readily extend this discussion to the Grassmann variable $\zeta = \xi + i\eta$. The real invariant here is $\bar{\zeta}\zeta = -i\eta\xi + i\xi\eta = 2i\xi\eta$, because of anticommutation. Here there exist three transformation operators which do not affect $\bar{\zeta}\zeta = i(\xi\eta - \eta\xi)$. They are

$$J_{\{\xi\eta\}} = \xi\frac{\partial}{\partial\xi} - \eta\frac{\partial}{\partial\eta}, \quad J_{\{\xi\xi\}} = \xi\frac{\partial}{\partial\eta} \quad \text{and} \quad J_{\{\eta\eta\}} = \eta\frac{\partial}{\partial\xi}$$

and they belong to the *symplectic* group Sp(2). If instead of ξ, η we use the complex Grassmann variables $\bar{\zeta}, \zeta$, with the invariant $Z = \bar{\zeta}\zeta$, then the symplectic generators are taken as

$$\left(\zeta\frac{\partial}{\partial\zeta} - \bar{\zeta}\frac{\partial}{\partial\bar{\zeta}}\right), \quad \zeta\frac{\partial}{\partial\bar{\zeta}}, \quad \bar{\zeta}\frac{\partial}{\partial\zeta}.$$

7.2 Isotopic shuffles

In fact the three Sp(2) generators above can be transmogrified into a spin algebra by forming the combinations:

$$S_3 \equiv iJ_{\{\xi\eta\}}/2, \quad S_1 \equiv (J_{\{\xi\xi\}} + J_{\{\eta\eta\}})/2, \quad S_2 \equiv -i(J_{\{\xi\xi\}} - J_{\{\eta\eta\}})/2$$

so $[S_1, S_2] = iS_3$ and cyclic. When they act on the doublet $\psi \equiv \begin{pmatrix} \xi \\ \eta \end{pmatrix}$ they correspond to a two-dimensional unitary group SU(2), such that $S_i \psi \equiv (\sigma_i/2)\psi/2$, where the σ are the famous Pauli matrices:

$$\sigma_1 \equiv \begin{pmatrix} 0 & 1 \\ 1 & 0 \end{pmatrix}, \quad \sigma_2 \equiv \begin{pmatrix} 0 & -i \\ i & 0 \end{pmatrix}, \quad \sigma_3 \equiv \begin{pmatrix} 1 & 0 \\ 0 & -1 \end{pmatrix}. \quad (7.1)$$

The reader will note that these spin matrices are traceless and hermitian and obey the simple algebraic rule: $\sigma_i \sigma_j = \delta_{ij}\mathcal{I} + i\epsilon_{ijk}\sigma_k$, where $\mathcal{I}$ is the 2D unit matrix and ϵ is the 3D 'Levi-Civita' tensor which is completely antisymmetric in all its indices, starting with $\epsilon_{123} = 1$, etc.

These matrices would normally operate on a two-component complex 'spinor' $\psi = (z_\uparrow, z_\downarrow)$ (for spin-up and spin-down) but one can well imagine them acting as 'isotopic rotations' on a complex nucleonic doublet such as (p, n) or on a leptonic pair (ν^0, e^-) associated with some fictional 'internal space'. A finite rotation θ about axis $\mathbf{n}$ of the doublet elements corresponds to an exponentiation

$$\psi \rightarrow \psi' = U\psi = \exp(i\Theta)\psi \equiv \exp\left(i\sigma_k n_k \frac{\theta}{2}\right)\psi$$

$$= \left[\mathcal{I}\cos\left(\frac{\theta}{2}\right) + i\sigma_k n_k \sin\left(\frac{\theta}{2}\right)\right]\psi,$$

where $\mathbf{n}$ is a unit three-dimensional vector. The astute reader will have noticed that I have exponentiated a 2×2 matrix $\Theta = \mathbf{n} \cdot \sigma\theta$. This may be new to some readers but is commonplace in mathematics and is no cause for concern: the result is given in the usual way by the series, $\exp(i\Theta) = \mathcal{I} + i\Theta + (i\Theta)^2/2! + (i\Theta)^3/3! + \cdots$. The reader may check that rotation matrix has unit determinant, which is what the letter S signifies in the isotopic group SU(2). The transformation U is said to be 'unitary' in as much as $U^\dagger = U^{-1}$.

When Θ is made x-dependent and we want the derivative $\partial\psi$ to transform the same as ψ we have to introduce a covariant derivative D which brings in a vector gauge field A to compensate for the change in location. Thus to make sure that the covariant derivative D does

its job, viz. $\psi D\psi \rightarrow \psi' D'\psi'$, we require $D = \partial - i\mathcal{A}$, provided that
the gauge field transforms according to the rule

$$\mathcal{A}' = U[\mathcal{A} + i\partial]U^{-1}. \qquad (7.2)$$

[For this 2D case it is not exceptionally hard to prove that the vector
A_i transforms as $A'_i = \cos^2\theta\, A_i + \sin 2\theta\, \epsilon_{ijk}n_j A_k + \sin^2\theta\, n_i n_j A_j - \partial\theta_i$,
in order that the location be irrelevant.]

The only matter which needs clearing up is how to ensure that
'curl' of the gauge field (in other words the analogue of the Maxwell
tensor F_{mn}) itself transforms correctly. Yang, Mills and Shaw [23]
found the solution to this non-Abelian problem: it is to construct
the 'covariant curl'

$$\mathcal{F}_{mn} = D_m\mathcal{A}_n - D_n\mathcal{A}_m = \partial_m\mathcal{A}_n - \partial_n\mathcal{A}_m - i[\mathcal{A}_m, \mathcal{A}_n]. \qquad (7.3)$$

We then find that $\mathcal{F}_{mn} \rightarrow \mathcal{F}'_{mn} = U\mathcal{F}_{mn}U^{-1}$, just like $\mathcal{A}$. Once
that is done, the Lagrangian for the SU(2) gauge field drops out as
$\mathcal{L}_A = -\mathcal{F}_{mn}\mathcal{F}^{mn}/4$. It implicitly incorporates a cubic and quartic
self-interactions. In turn this complicates the perturbative treatment
of the gauge field through higher powers of A. And when the quantisation of the gauge field is considered — which means fixing a
gauge — the complications get worse because of the need to introduce intermediate ghost fields for ensuring unitarity (or probability
conservation).

7.3   Colour changes

When we come to the matter of transforming colour, we are dealing
with the group SU(3) rather than SU(2). Instead of the three 2×2
Pauli matrices σ_k we must substitute the eight 3×3 Gell-Mann [24]
matrices λ_i where $i = 1, \ldots, 8$. These λ are all hermitian and traceless
but obey a rather more complicated algebra, summarised by the rule
$\lambda_i\lambda_j = 2(d_{ijk} + if_{ijk})\lambda_k$; the f_{ijk} are the 'structure constants' of the
SU(3) Lie algebra and are fully antisymmetric in their indices and
are similar to the Levi-Civita ϵ_{ijk} of SU(2). Actually we can avoid
getting enmeshed in detailed algebra and bypass studying these constants comprehensively by constructing the 3×3 chromodynamic

matrix $\mathcal{A} \equiv \lambda_j A_j/2$. Acting with a 3×3 unitary matrix $U(\Theta)$ on a triplet of colours shuffles them around and rotates the gauge field according to Eq. (7.2). This leaves invariant the scalar combination $Z_c \equiv \zeta^{\bar{\mu}}\zeta^{\mu} = \zeta^{\bar{1}}\zeta^1 + \zeta^{\bar{2}}\zeta^2 + \zeta^{\bar{3}}\zeta^3$, where the chromic triplet is $(\zeta^1, \zeta^2, \zeta^3)$.

The gluon interactions with the source fields ψ and with themselves mimic the SU(2) case. We still maintain SU(3) gauge invariance through the two sets of terms:

$$\mathcal{L} = \bar{\psi}[i\gamma \cdot (\partial - iA) - m]\psi - \mathcal{F}_{kl}\mathcal{F}^{kl}/4 \qquad (7.4)$$

where the matrices $\mathcal{A}, \mathcal{F}$ are 3×3 in size. It should be perfectly obvious to the reader now how the whole procedure can be extended to greater numbers of properties. The only point left to emphasise is that Eq. (7.4) refers to massless gluons — a mass term such as $A \cdot A$ potentially destroys the local gauge symmetry because it changes with phase and therefore cannot be entertained! It indicates that massive gauge bosons are excluded. On the contrary the weak interaction bosons are quite massive; so we have before us a big dilemma.

That is how the subject stood in the early 1970s until the discovery/realisation of spontaneous breakdown of symmetry.

7.4 Grand connections

The idea of grand unification is to combine leptons and quarks into some larger grouping and allow transitions between them; so at the very least it means somehow uniting the five complex properties ζ^0 to ζ^4, disregarding handedness. This idea is potentially dangerous because any shuffling between baryons and leptons might lead to nucleon decay into leptons through intermediate particles that connect with both. While it is true that we would need to get all three quarks to decay successively into the lepton sector this might possibly result in a very long proton lifetime (much longer than the present age of the universe). However there are an awful lot of nucleons in the universe; thus even if a miniscule proportion disintegrate, they should be detectable. So far there is no evidence whatsoever for proton decay!

I shall return to the matter of grand unification and provide some details about their consequences in later chapters, including incorporation of the Standard Model. But I need at this stage to point out that having five complex ζ properties or 10 real (ξ, η) leads one naturally into grand SU(5), SO(10) and Sp(10) groups. It also prompts us to consider the idea of ζ combinations which have mixtures of leptonic and hadronic characteristics, going under the name of 'leptoquarks'. The most natural amongst these would be gauge bosons that latch electrons (or neutrinos) to quarks, which permit perilous hadron–lepton transmutation. Other gauge groups have been suggested, such as the exceptional group E(8), but they have not gained wide acceptance nor great prominence as yet, but may do so in future.

Chapter 8

Dreams of Unification

During the last few years of his life, Einstein strove to unify electromagnetism with gravity in a geometrical way. He was much taken by the idea of Kaluza (further developed by Klein) that a fifth bosonic [25] coordinate x^5 added to the four spacetime coordinates (x^0, x^1, x^2, x^3) will achieve that end, but Einstein was never completely satisfied with the idea, as I shall sketch. Einstein knew that he had to achieve that unification first before turning to the other forces of nature, of which he was well aware, but which he studiously ignored initially.

Einstein's disciples are legion and many others have pursued those unification aims more grandiosely: to unify *all* the forces of nature geometrically and attain a 'theory of everything'. Needless to say that goal has proved elusive, with quantisation providing an additional obstacle. In what follows I will try to sketch the main ideas that have been promulgated and where they have come to grief or at least not reached fruition. Some of the topics are really outside my area of expertise, but I will attempt to convey the general ideas and summarise why they seem to have faltered. I will try to be fair-minded and not over-critical. It is possible, even if appears unlikely today, that some of the concepts may eventually blossom. Inevitably, practitioners of some of these specialised topics may not accord with my perception about the state of the union; if they have a bone to pick, I would welcome hearing from them so that, if ever a second edition of this monograph comes into existence, I can modify the record.

Fig. 8.1   Theodor Kaluza.

Fig. 8.2   Oskar Klein.

8.1   Electric dreams

If there is a fifth coordinate — call it x^5 — where on Earth is it hiding? Klein suggested that it might be miniscule and correspond to a closed circle. Kaluza originally envisaged the metric element g_{m5}, from spacetime to a spatial fifth dimension, as corresponding to the electromagnetic field but made the assumption that it did not depend on x^5. That is what stuck in the craw. Klein clarified the assumption by visualising x^5 as running along a circumference of a closed tube, the new coordinate thus being curled up into a *very tiny circle* which he estimated to be around 10^{-33} m — a Planckian scale associated with Newton's constant G_N. The dependence of fields on this new coordinate can then be summarised by the mode expansion,

$$\Phi(x, x^5) = \sum_{n=-\infty}^{\infty} \phi_n(x) \exp(inx^5/R); \quad x^5 \in [0, 2\pi R). \qquad (8.1)$$

Such mode expansions apply to the metric and in particular to the vector sector $g_{m5}(x, x^5)$ and the scalar sector $g_{55}(x, x^5)$. Klein regarded the electromagnetic field $A_m(x)$ as the 'zero-mode' $g_{m5\,0}(x)$ originally introduced by Kaluza.

The first question then arises: what about the other modes? Well, the extended Klein–Gordon equation, $(\Box - \partial_5{}^2)\Phi = 0$ will produce a series of mode-dependent mass terms $m_n = n/R$ via Eq. (8.1). So indeed the zero mode ($n = 0$) is massless as befits a photon. However the other modes ($n \neq 0$) are inversely proportional to the radius of the circle. This is not unexpected but is something of a worry as there are an infinite number of them! As already stated, the way out is to assume that the radius R is exceedingly tiny so that the higher modes are extremely massive and way out of experimental reach. I believe this was the first aspect that concerned Einstein.

Einstein's second and more serious concern was that even if one presumes that the extra metric elements do not involve x^5, they still depend on 4D spacetime x. In particular the g_{55} sector introduces a scalar field and Einstein was not enamoured of a mixed scalar–tensor theory. As a matter of fact, such affliction persists in all geometrical models of elementary fields which extend spacetime by extra bosonic coordinates like x^5. Furthermore the scenario where these additional tiny coordinates curl up into 'Calabi–Yau' manifolds represent a nightmare rather than a pleasurable dream.

This has not deterred physicists from contemplating, indeed welcoming, the complications arising from extra coordinates. They have extended Klein's picture from the electromagnetic group U(1) to isotopic SU(2) and QCD SU(3); it all works delightfully out provided the familiar gauge fields for those groups are restricted to zero modes in those superfluous coordinates. Taking the extra scalars to be constants leads to further miracles, in so far as the Lagrangian for gravity and the gauge field drop out, with the field equations matching the general relativistic ones for the curvature tensor and the generalised Maxwell fields. Others have taken Klein's concept more seriously and searched for signals of the various supernumerary radii; so far any experimental detection of not too small dimensions R or supermassive states has gone unrewarded. For a modern treatment of Kaluza–Klein theory, I recommend the book by Appelquist, Chodos and Freund [26]; it repays further reading handsomely.

8.2   Grand designs

The gauge groups for the strong interactions and the electroweak interactions [27], namely SU(3) and $SU(2)_L \otimes U(1)_Y$ are distinct and have little to do with one another: they are taken as 'direct product' groups. It takes some chutzpah and real gumption to link them more intimately, but that is what Georgi and Glashow [28] and Pati and Salam [29] did in the early 1970s. They tried in different ways to combine the quarks and leptons into a larger multiplet as part of some grander group.

Georgi and Glashow envisaged an SU(5) group with each generation of quarks and leptons fitted into a $5 \oplus \overline{10}$ left-handed representation. As the antiparticle or conjugate states are included, the scheme also incorporates the right-handed states. At that time neutrinos were thought to be massless and the right-handed versions were thought to be non-existent; however since then neutrinos have been found to carry a little mass so ν_{right} must exist and be included as a singlet of SU(5). See Figure 8.3.

On the other hand, Pati and Salam regarded leptons as a fourth colour (lilac) beside the three strong QCD colours (red, green, blue). So they had doublets falling into an SU(4) extended colour group with a direct product of an isotopic SU(2) (of both parities). Figures 8.3 and 8.4 below show these respective choices. Of course both marriages are perfectly allowed provided that leptons and quarks are not tied too closely. But the proponents of these schemes do link them via gauge fields at the very least (not to mention 'leptoquarks'); otherwise there would be no point in such matchmaking. A big danger then looms: it might become possible for quark mixtures (such as the proton) to decay into leptons, leading to the instability of baryonic matter. If the process is extremely slow then one can escape such a dreadful outcome. Originally the

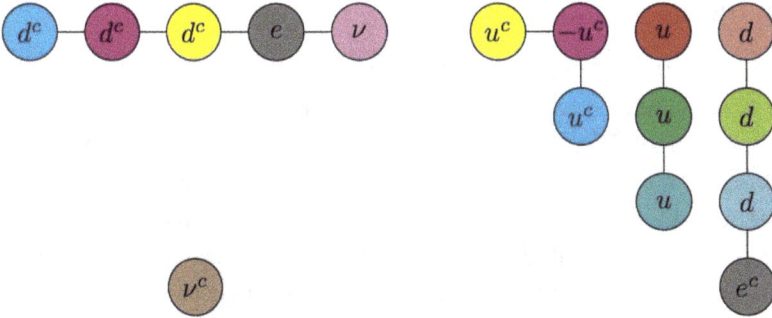

Fig. 8.3   Georgi–Glashow multiplets of $1 \oplus 5 \oplus 10$.

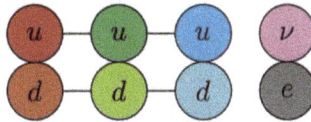

Fig. 8.4   Pati–Salam unified multiplet, with leptons as the fourth color.

half-life of the nucleon was estimated to be of the order of 10^{32} years; this turned out to be unacceptably low because it is easily verifiable by looking at matter decay — there are an awful lot of nucleons around us! The experimental lower limit on the lifetime is now set at about 10^{35} years so this greatly restricts the masses of the gauge fields and scalar leptoquarks that can mediate such matter disintegrations. Today these grand marriages have ended up in separation, if not uncontested divorce, with one party not even making a court appearance.

To complete this rapid survey I would be remiss in not mentioning that the group SO(10) is an even more *grandiose* unification scheme. The Lie group encompasses both SU(5) and SU(4)⊗SU(2). It was originally promoted by Fritzsch and Minkowski [30]; the quarks plus leptons fit nicely into a 16-fold 'spinorial' representation of the group. The SO(10) model still enjoys comparative popularity though it is of course subject to the same comments as Pati–Salam and Georgi–Glashow grand models.

8.3   Super visions

The mid-1970s heralded a period of great optimism in particle physics. It was sparked by the discovery of fermion–boson super-symmetry (SUSY), which came 'out of the blue' from the researches of Volkov and Akulov [31], Wess and Zumino [32] and Salam and Strathdee [33]. It is based on transformations between the two types of statistics by the introduction of fermionic parameters. The enthusiasm for this idea was the promise that it escaped the no-go theorems of O'Raifearteagh [34] and Coleman and Mandula [35] for enlarging the relativistic Poincaré group non-trivially. The concept corresponds to 'square-rooting' the group of translations, mimicking the way that Dirac 'square-rooted' the Klein–Gordon equation and thereby linearised it.

SUSY rests on the basic anticommutation relation between supersymmetric operators Q carrying spinorial indices, like the electron:

$$\{Q_\alpha, Q_\beta\} = (\gamma_m C)_{\alpha\beta} P^m; \; C = \text{charge conj. matrix},$$

$$P^m = \text{momentum}, \tag{8.2}$$

accompanying those of the Poincaré group for special relativity. Associated with these new 'supertranslation' operators Q_α are parameters θ_α of some internal spinorial space. Superfields Φ then become functions of spacetime x plus these new parameters θ; because these θ carry 4 spin components the expansions in θ terminate at the fourth power and produce spins 0 to 1/2 to 1 until they peter out; taking the θ as real Majorana spinors (so $\psi_\alpha = \psi_\alpha{}^c = C_{\alpha\beta}\bar{\psi}^\beta$), the expansion reads

$$\Phi(x,\theta) = S(x) + \bar{\theta}^\alpha \psi_\alpha(x) + \bar{\theta}^\alpha \bar{\theta}^\beta (\sigma^{mn})_{\alpha\beta} F^{mn}(x)/2$$

$$+ \epsilon_{\alpha\beta\gamma\delta} \bar{\theta}^\alpha \bar{\theta}^\beta \bar{\theta}^\gamma \left(\bar{\chi}^\delta(x)/3! + \bar{\theta}^\delta P(x)/4! \right). \tag{8.3}$$

Because Φ is overall bosonic, so are the even coefficients S, F_{mn} and P. However the odd coefficients ψ, χ are fermionic. This should be familiar as the 1+4+6+4+1 sequence of numbers from the binomial expansion of $(1+x)^4$ in powers of x. Thus Φ splits into 8 bosonic terms and 8 fermionic ones. It is a general feature of such supersymmetric expansions in auxiliary fermionic variables like θ and is one of the most attractive ones; for it has the potential of taming commonly occurring infinities due to the cancellation between bosonic and fermionic quantum loop corrections.

Practitioners of SUSY for the real world always have matching pairs of supersymmetric partners: gluons and 'gluinos', quarks and 'squarks', photons and 'photinos', electrons and 'selectrons', etc. Experimentalists have frantically searched for signals from the superpartners (which might potentially be heavy) but so far without any discernible success. Despondency has now set in that many believe they will never be found and probably do not exist. If so it spells the end of a really brilliant concept. Even so, theorists have extended SUSY in different ways into new domains in highly imaginative and original ways. Foremost among such extensions have been (i) enlarging the idea to supergravity [36] (when the graviton is accompanied by its gravitino spin 3/2 superpartner), (ii) tackling grand unified versions in the hope that this will alleviate matter lifetime concerns, and (iii) suggesting internal 'horizontal' symmetries, by invoking N-extended θ parameters. The latter extension is a spinorial variant of my own attempts later on with scalar anticommuting parameters, but it encounters serious spin proliferation. Having said that, the $N = 4$ and $N = 8$ versions possess beautiful properties and therefore many acolytes.

8.4 Stringent views

The old picture of point particles had to undergo revision with the advent of one-dimensional strings and 'branes' (extended structures) in a higher dimensional space. Stringy ideas themselves had a curious birth. In the late 1960s S-matrix self-consistent scattering processes were all the rage; the idea was that amplitudes could either be expressed as a sum of poles/resonances in one channel or equivalently in a crossed channel, depending on the way one views the corresponding Feynman picture for the process. This culminated in Veneziano's beta-function characterisation,[1] which encapsulated the

---

[1]See Section A.6 to follow the mathematics behind this.

concept, and led to the question: what fundamental dynamics could produce such amplitudes?

Nambu and Goto [37] were among the first to produce a bosonic action that could do the trick by adding an extra spacelike stringy coordinate σ to proper time τ and associating the action with the area traced out by the extended coordinate $X(\tau, \sigma)$. The scheme introduced a new fundamental scale, characterising the 'string tension'. Polyakov [38], Deser and Zumino [39] elaborated upon and tidied up the formulation by likening it to general relativity. It was discovered that the purely bosonic string was only consistent if it existed in 26 dimensions, otherwise unwanted 'tachyons' would result — an unpleasant feature. In spite of this setback, the spinning string came soon after, through the work of Ramond [40], Neveu and Schwarz [41] who incorporated fermions; this was a good thing because consistency was established in a total of 10 critical dimensions, a big improvement over 26! With the incorporation of fermionic vibrations the bosonic string was transmogrified into a 'superstring'.

The subject lay dormant by mainstream theorists until the mid-eighties when it was realised that string theories could in principle, if not in practice, characterise all the known elementary particles and their interactions; most importantly they include the graviton as one of the fundamental vibrations. Further Green and Schwarz [42] showed that they were miraculously free of anomalies. Strings can be open or closed — with differing boundary conditions — and five types of consistent string models are now recognised.

The next step to progress was achieved when Witten [43] recognised that these five models were different manifestations of an 11-dimensional picture, called M-theory, through various equivalences or dualities. The subject got further complicated when Polchinski [44] conceived of structures more extended than 1D strings, which are called 'branes'. Because of the many possibilities for the 6 or 7 dimensions extra to spacetime, for curling up, confusion reigns at present. It has to be admitted that in spite of the elegant mathematical complexities and technicalities of string theory there is no single prediction that can be regarded as unequivocal evidence for the idea. Mother Nature seems to have spurned the string world view. In fact the multitude of 'landscapes' conjured up by compactified string theory have landed us precisely nowhere, which is extremely disappointing: a lot of persons have invested much time and effort into that field and are not willing to let go.

8.5 Dark thoughts

Paradoxically, through painstaking investigations by galactic astronomers over the last century, it has come to light that some large scale dark forces are operating in the universe. Dark matter and dark energy are two aspects of this. I will focus on the former as the latter is tied intimately to cosmological models. By its very nature, dark matter has remained hidden from view and therefore escaped particle physicists' notice until the last couple of decades. There are five indicators and they either point to the existence of dark matter or they require large scale modifications to Newton's gravitational force law. Here is a brief rundown of the current state of affairs.

By studying the binding motion of galaxies within a galactic cluster (the Coma cluster), Zwicky concluded that they could not hold together without some invisible extra matter being present to provide the gravitational attraction, without radiating much if any light. The second clue came from studying of the mass distribution and motion within our own Milky Way where evolution seemed to suggest that our star distribution would look very different from what it is today without some considerable hidden mass component to shape the present distribution. This picture was reinforced by the investigations of Ford and Rubin on the motion of hydrogen clouds in the Andromeda galaxy. Instead of tailing off from the centre rapidly, as Newtonian observations of the visible stars would suggest, the gas motion steadied out with distance, as if some additional component of the galaxy was holding it together. The fourth clue comes by detailed studies of the 'cosmic microwave background' and the small variations of temperature in angular extent, which indicate that normal matter was and is dominated by dark matter at some stage during the universe's expansion/cooling. The last clue comes from gravitational lensing of distant very bright quasars, showing there is some additional stuff between the quasar and ourselves which produces the multiple images.

From these experimental hints one must either (i) accept that intergalactic space is permeated by a substantial amount of non-interacting stuff (apart from its gravitational attraction) or (ii) that Newtonian dynamics is deficient at galactic scales. Most researchers have opted for case (i) and instigated searches for dark matter by looking carefully at processes where there is missing energy or momentum in the visible stuff; they ascribe that phenomenon to dark matter silently taking off the excess. A small but devoted coterie is open to case (ii) and they investigate Modified Newtonian Dynamics (MOND), taking care to not disturb GR too much, on which Newtonian dynamics relies; their efforts should not be discounted because the subject is still as murky as ever.

Assuming that (i) is a more likely scenario, the question arises as to what the dark materials consist of; we think we can feel it but we cannot see it and this tells us a bit, namely that the hidden stuff is intrinsically dark or too faint to see, so that its electromagnetic interactions are negligible. Dark matter candidates are:-

- MACHOs (MAssive Compact Halo Objects): These are astronomical objects that have not lit up, like failed stars or heavy planets. A survey concentrating on the region of the Magellanic Cloud has shown no sign of these through possible gravitational distortion of light caused by MACHO obstruction.
- WIMPs (Weakly Interacting Massive Particles): Models of families of particles have thrown up states which interact weakly with the usual normal collection but are extremely massive. All accelerator searches for these have failed so far and the Large Underground Xenon detector (LUX) has not seen them either.
- FILPs (Feebly Interacting Light Particles): Foremost amongst these are the axions, which were invoked by Peccei and Quinn [45] to show how and why strong PC conservation holds. These hypothetical axions have no spin and charge and are thought to be very light indeed but will interact with photons to produce two photons in the form of radio waves. There are no signs of them. Some supersymmetric models predict the existence of φ particles (of order MeV) and gluino partners of the gluon, again without

any experimental success; similarly for neutralinos and gravitinos that come with SUSY.

- Neutrinos are known to be massive via their peregrinating trans-mutations, and their masses are of the order 0.1 eV or less. Because they are so elusive due to their weak interactions they might constitute at least some of the putative dark matter. If they were 'cold' or slow moving (meaning they have low energy) then it has been suggested that they would cause too much clumping in the early universe which would be detectable in the cosmic microwave background — and is definitely not seen. The current favourites are hot neutrinos which might account for a fraction of the dark matter density. Where the remaining dark density comes from is still something of a mystery.

8.6 Flightless fancies

Particle physics finds itself in something of an impasse today. Imaginative proposals such as those I covered above have not panned out, in as much as the Standard Model seems to account for pretty well all the reactions observed without any further elaboration or modification. Yet some mysteries about the Standard Model demand demystification. For instance:

- Are the current masses/mixings of quarks and leptons really just independent parameters (or equivalently their Yukawa couplings to the Higgs field)? May they be related in some way? Empirical models have been espoused in which there is just such a connection but they seem to lack a fundamental basis.
- What is the reason for triplication of generations? After all, the world would not be very different if there was just one generation. Are there other generations lurking in the background, accessible at higher energy? Do there exist other states that do not form part of the standard picture?
- With the failure of SUSY, are there other frameworks into which the Standard Model finds a natural fit? Will grand unification come of age, avoiding matter decay?

- Do cosmological ideas have an impact on force unification? For example, does the AdS/CFT correspondence have anything to say about that, or will the inflaton effects in the early evolution of the universe find a comfortable spot in the scheme of things? Perhaps the holographic principle may offer a new picture for unification?
- The weak bosons discovered thus far have left-handed interactions. Do right-handed weak bosons exist and await discovery? The neutrino has surprised us by possessing mass and therefore comes with both parities so perhaps the same applies to W and Z.
- What role does 'naturalness', like anomaly cancellation, play in constraining the values of couplings? So far the former has had little or no impact as far as I am aware.
- Does the anthropic principle — more on that later — circumscribe the free parameters of the observable universe and does it advance the science of particle physics?

Small wonder then that these questions have had theorists exploring many different byways in the hope that some mysteries might reveal themselves. I will try to summarise some of these concepts to the best of my knowledge.

8.6.1 *Horizontal projections*

The established Standard Model picture has *three* generations of leptons plus triply coloured quarks, but offers no explanation for the repetitions. If we arrange them in a row:

$$\begin{pmatrix} e \\ \nu_e \end{pmatrix} \begin{pmatrix} \mu \\ \nu_\mu \end{pmatrix} \begin{pmatrix} \tau \\ \nu_\tau \end{pmatrix} \quad \text{and} \quad \begin{pmatrix} u \\ d \end{pmatrix} \begin{pmatrix} c \\ s \end{pmatrix} \begin{pmatrix} t \\ b \end{pmatrix}$$

we can envisage a classification group (blind to colour) that shuffles the elements in a horizontal sense.[2]

As there is no observational evidence for $e - \mu - \tau$ transitions one has to be very careful not to allow such transmutations when

---

[2]It should be pointed out that the mass eigenstates of the lower elements are mixtures of the flavour states, but the upper elements are not.

developing a gauge theory that goes across horizontally. Therefore although one can conceive of triplet representations of some horizontal group, such as SU(2) or SU(3) say, it is very dangerous to gauge all the fields that accompany the full gauge group. These horizontal groups can serve as classification aids certainly but must be treated with care; they help to organise the generations, in the same way that the eightfold way of Gell-Mann and Neeman [24] was used in the past to characterise the flavour triplet (u, d, s).

It is necessary to constrain the horizontal gauge groups to the subgroups $U(1)_e \otimes U(1)_\mu \otimes U(1)_\tau$ thereby preventing unwanted transitions between the charged leptons. In reality this does not advance us very far because it effectively treats each generation differently, with its own set of quantum numbers; it has not progressed the field very much.

8.6.2 *Fruitless loops*

We are all well aware that at Planck energies or distance scales $(10^{-35}\,\text{m})$ physics must undergo drastic revision, for at such scales, gravity becomes as strong as the other forces and the concept of a point location becomes suspect: the quantum measurement fluctuations will place the notion of a coordinate in jeopardy and spacetime might spread out into some kind of foam. A small but dedicated group of researchers have investigated the idea that spacetime itself should be quantised, based upon an approach initiated by Abhay Ashtekar. While this sort of research has little to say about force unification it has its small band of devotees so I will briefly sketch out the theory of loop quantum gravity (LQG) despite its problems.

The idea is to make the coordinates non-commutative leading to a new kind of non-commutative geometry. It is assumed that there is a minimum size to space, governed by the Planck scale. In that approach there is a minimal area and volume and a set of quantum states are built upon these minimal structures. Spacetime is viewed as a series of intersecting loops of gravitational field at those scales. The connections between adjacent mini-volumes can be interpreted as connected graphs and lead to a 'spin network'. Time evolution is conceived as the way the spin network evolves. It is not known whether classical general relativity emerges at larger scales and there are many unanswered questions which are far from being resolved. They have not yet borne fruit which explains the cheeky title of this section. Still the brave band of LQG afficionados are undeterred and continue to soldier on. For those readers who are interested in pursuing this topic I suggest reading the articles by Rovelli [46] and Smolin [47].

8.6.3   *Principled perceptions*

In trying to limit the number of possible speculative research directions, a set of principles are often enunciated which border on philosophy. Foremost among these is the *anthropic principle*. As originally stated by Carter [48], who leant on previous remarks by Dicke, the principle comes into varieties: the weak and the strong. The weak principle states that as humans we are positioned in just the right spacetime location of the universe to act as observers; whereas the strong version says that the basic parameters of physics are what they are in order to give rise to humans who may then observe it. An extension of the principle by Barrow and Tipler [49] enlarges the concept to carbon based life forms such that observers are required to bring the universe into existence among many possible other universes. As far as I am aware, the only new prediction that emerges from the anthropic principle is the value of the cosmological constant by Weinberg; its ridiculously small value is 120 orders of magnitude smaller than what is expected from particle physics; but Weinberg [50] shows that it is what is needed for the universe to develop in the way it does to enable the formation and evolution of galaxies.

Beauty has played a significant role in shaping elegant theories and getting them accepted by mainstream physicists; it has also governed the search for theoretical generalisations. An idea that dimensionless ratios of fundamental constants of the same ilk should be of order 1 has recently taken root; the archetypical example is the mass and vacuum expectation values of the Higgs boson, which theoretically should be many orders of magnitude larger (based on quadratic infinities from quantum loops) — the actual scale of about 200 GeV seems 'unnatural' and requires 'fine-tuning'. To explain its seemingly unnatural value, supersymmetry has been invoked, since the fermion loops can cancel out the bosonic loops; to no avail as it happens, since no quark and lepton SUSY partners have revealed themselves. There are plenty of so-called unnatural ratios, like the ratio of the top quark mass to those of neutrinos which can be as much as 10^{10}! Not to mention the worst scale ratio of all: the cosmological constant scale to the weak scale. In short there is now little confidence in the truth of the *naturalness principle* and we may have to accept that some ratios are fine-tuned. Physics can be ugly and many dimensionless ratios can be absurdly large, as Hossenfelder [51] has stressed and deplored.

Another principle has recently been making the rounds. It is based on the relationship between anti-de Sitter spaces and conformal (scaleless) theories and has led to the *holographic principle* as promoted by 't Hooft, Thorn and Susskind [52–54]. It proposes that the four-dimensional world can be viewed as a boundary surface of a higher five-dimensional space, encoding all the properties we know and love as some kind of hologram. It sprang into being by the discovery that information lost as one crosses a black hole event horizon is held in the surface fluctuations of the horizon, thereby resolving a thermodynamic paradox. Readers should consult the article by Maldacena [55] if they wish to explore the idea further.

We now come to unprincipled speculations. Because it is unclear how particle physics can get out of its present doldrums, a scattergun approach has been taken by a number of people. All sorts of extra particle multiplets with unconventional interactions and new symmetries have been invoked to explain dark matter and a tiny number of experimental results[3] that seem to contradict Standard Model expectations. Leading contenders are the states predicted by grand

---

[3]Rare B-decays into K^* and D^* seem to deviate from Standard Model predictions, corresponding to the quark decays: $b \to s\ell\bar{\ell}$ or $b \to c\tau\nu$.

unified models, new Higgs fields and right-handed gauge bosons. We should not discount these apparently random explorations; a lucky shot in the dark may be just what one needs to make truly significant progress; the clues that have been pursued so far have led us down the garden path.

Chapter 9

Dualities

Before going further I ought to mention other physicists who have shared aspects of the philosophy behind this monograph. Foremost amongst these has been Casalbuoni and his collaborators [56], who have also considered blends of Fermi oscillators, instead of using anti-commuting coordinates, but effectively with the same sort of algebra. They have attempted to obtain a unified description of quarks and leptons just by using three coloured creation operators plus an additional colourless one. To get generations they include a further set m of Fermi oscillators and are led to Lie groups of type $O(10) \times O(2m)$. However they did not go so far as geometrising their work nor unifying it with gravity.

Let me now outline some of my own attempts at *geometrical unification of forces*; in the process we might shed some light on the 'generation problem' — why there exist repetitions of quarks and leptons. As a byproduct it will be necessary to introduce a scale to tie the dimensionless properties ζ to spatial x coordinates; that scale will be intimately connected with Newton's gravitational constant. The motivation is that only by going back to basics can we make progress along the lines mentioned in Section 6.3. Thus we will be introducing a number (effectively 5 will suffice) of scalar anticommuting complex variables $\zeta^0, \zeta^1, \ldots$, which represent the properties or attributes of a system. They serve to characterise the 'what' of an event in addition to the usual 'when–where' of the occasion.

9.1 Grassmann duals

Quantum probabilities in their simplest formulation are determined by the action principle: the action, being the integral of the Lagrangian (which embodies the dynamics) over all coordinates, has to be minimised, leading to the Euler–Lagrange equations of motion. In our case we must include the property variables ζ as well as the spacetime coordinates x, thereby summing over all possible events. In the past we had to collect fields and events separately and add them together as a series of interaction Lagrangian terms, whereas now the inclusion of ζ should automatically cover all events upon integrating over properties. In this chapter the focus will be on the nature of property expansions and whether we can impose symmetries that minimise the number of summations.

Let us look at the U(1) case first where there is a single ζ. Duality is of very little consequence here but there are subsidiary issues which will play a role later. If we were to expand a general *bosonic* real function $F(x, \zeta, \bar{\zeta})$ in powers of the ζ we would encounter the polynomial

$$F(x, \zeta, \bar{\zeta}) = A(x) + \bar{\zeta}\psi(x) + \psi^c(x)\zeta + \bar{\zeta}\zeta B(x); \quad F = F^\dagger, \text{ if real.}$$

Because the ζ anticommute, we deduce that the components A and B are bosonic while ψ, ψ^c are fermionic. It makes sense to separate F into even and odd parts because they behave differently under Lorentz transformations. Set the bosonic and fermionic superfields to be

$$\Phi(x, \zeta, \bar{\zeta}) = A + \bar{\zeta}\zeta B; \quad \Psi(x, \zeta, \bar{\zeta}) = \bar{\zeta}\psi(x) + \psi^c(x)\zeta.$$

After making this manoeuvre, A and B are scalar like Φ and carry no spin labels, but ψ is fermionic and we need to attach a spinorial index to it as well as its progenitor Ψ. We must pay homage to the spin-statistics connection, so separating Φ from Ψ does duty to that theorem.

Now to the matter of duality. If we lay the F expansion terms in the form of a 2×2 checkerboard so that ζ powers run to the right (from 0 to just 1 in this trivial case) and powers of $\bar{\zeta}$ run downwards, the fermions fall on the black squares and bosons find their places on the white squares. When we reflect about the main diagonal we are changing the property into its conjugate, so we see that A and B are self-conjugate (or real) fields, whereas ψ and ψ^c are charge conjugates of one another. So far, so obvious.

But we can imagine another transformation which corresponds to reflection about the cross-diagonal; it takes bosonic components into one another and likewise for fermionic components. (This is always true, irrespective of how many independent properties there are.) We can define a duality ζ-operation for such a cross reflection, symbolised by a $^\times$ superscript, whereby

$$1^\times = \bar\zeta\zeta, \quad (\bar\zeta\zeta)^\times = 1 \quad \text{and} \quad \zeta^\times = \zeta, \quad \bar\zeta^\times = \bar\zeta.$$

What is the point of that, you may be thinking; well, if one imposes self-duality on $F(x, \zeta, \bar\zeta)$ then we can identify cross-reflected states as being one and the same, and so cut down on the number of independent components. With just one property this is very simple. Define $Z \equiv \bar\zeta\zeta$, as it occurs very frequently. For

- selfdual F one has $A = B$, so $\Phi = (1 + Z)A/2$ and $\Psi \neq 0$, while
- anti-selfdual F has $\Phi = (1 - Z)A/2$ and $\Psi = 0$.

These duality symmetries have repercussions when we come to constructing Lagrangians because they simplify the whole process.[1]

Assuming Grassmann selfduality, we find that

$$\int d\zeta d\bar\zeta \; \Phi^2 = A^2/2, \quad \int d\zeta d\bar\zeta \; \bar\Psi\Psi = \bar\psi\psi/2, \tag{9.1}$$

provided that the adjoint of $\Psi(\zeta, \bar\zeta)$ is defined to be

$$\bar\Psi \equiv \bar\zeta\bar\psi^c - \bar\psi\zeta.$$

We can make life easier for ourselves by halving the Ψ expansions by dropping the charge conjugates ψ^c and its associated ζ; such terms only serve to double the results and that is what I will presently do.

---

[1] The duality concept can be extended to SUSY and can produce easier Lagrangians without necessitating the use of the D derivative in that particular scenario, quite apart from reducing the number of θ-expansion coefficients in superfields.

9.2　SU(2) dual states

For the case of two leptonic properties, like (ν^0, ℓ^-), duality becomes a more meaningful and useful concept. Given the corresponding labelling (ζ^0, ζ^4), their conjugates $(\zeta^{\bar 0}, \zeta^{\bar 4})$ and the invariant $Z = \zeta^{\bar\mu}\zeta^\mu = \zeta^{\bar 0}\zeta^0 + \zeta^{\bar 4}\zeta^4$, the duality operation acts as follows[2]:

$$1^\times = Z^2/2, \quad (\zeta^\mu)^\times = Z\zeta^\mu, \quad (\zeta^0\zeta^4)^\times = -(\zeta^0\zeta^4), \text{ etc.} \quad (9.2)$$

From these duality rules we can construct bosonic Φ and fermionic Ψ superfields with specific duality symmetries:

$$\text{Selfdual } \Phi = (1 + Z^2/2)\phi + \zeta^{\bar\mu}\rho^{\mu\bar\nu}\zeta^\nu;$$

$$\rho = \text{isotriplet}, \quad \phi = \text{isosinglet},$$

$$\text{Anti-selfdual } \Phi = \phi(1 - Z^2/2) + \zeta^{\bar 4}\zeta^{\bar 0}\varphi + \bar\varphi\zeta^0\zeta^4;$$

$$\phi = \text{isosinglet}, \quad \varphi = \text{complex isosinglet}. \quad (9.3)$$

$$\text{Selfdual } \Psi = (\zeta^{\bar 0}\nu + \zeta^{\bar 4}\ell)(1 + Z);$$

$$\text{Anti-selfdual } \Psi = (\zeta^{\bar 0}\nu + \zeta^{\bar 4}\ell)(1 - Z). \quad (9.4)$$

Because we made no distinction between left and right-handed property at this stage, and we have ignored chromicity altogether, there is just one generation of leptons to consider at this point whether or not we impose selfduality or anti-selfduality on superfields. Shortly we shall be distinguishing between left and right weak isospin and come across the corresponding invariants, $\overline{Z}_L$ and Z_R and their powers as we will be tackling electroweak theory.

9.3　SU(3) dual states

With colour, the plot thickens and duality possibilities open up. Again we can lay out the states

$$F = \sum_{r,s=0,\dots,3} (\zeta)^r (\bar\zeta)^s f_{rs}$$

on a 4×4 checkerboard labelled by the powers r, s as shown in Figure 9.1. We can group state combinations as to whether they are

---

[2]See Reference [57] for a more detailed explanation.

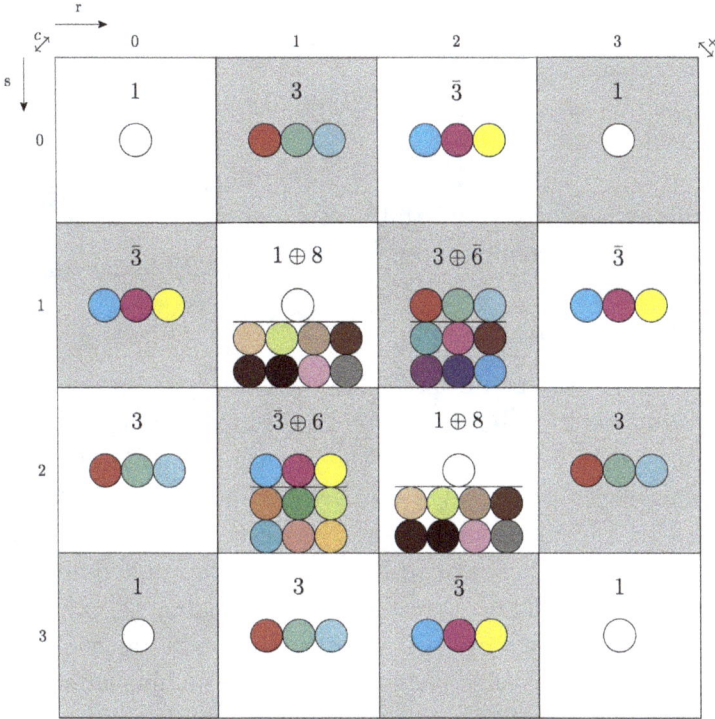

Fig. 9.1 Coloured multiplets. Bosons fall in the white squares and fermions in the grey ones. The SU(3) multiplets are labelled by their sizes and are represented by the coloured circles. Conjugation (c) corresponds to reflection about the main diagonal while the duality operation ($\times$) corresponds to reflection about the cross-diagonal. Notice that duality symmetry can be imposed without affecting the character of the multiplets.

selfdual or anti-selfdual as follows using the chromic singlet abbreviation $Z_c = \zeta^{\bar{1}}\zeta^1 + \zeta^{\bar{2}}\zeta^2 + \zeta^{\bar{3}}\zeta^3$:

$$\text{Selfdual } \Phi = (1 + Z_c^3/6)\phi + Z_c(1 + Z_c/2)\varphi$$
$$+ (\bar{\chi}\zeta\zeta + \bar{\zeta}\bar{\zeta}\chi)(1 + Z_c) + \bar{\zeta}\rho\zeta(1 - Z_c);$$
$$\rho \text{ octet}, \chi \text{ sextet}, \phi, \varphi \text{ singlets}; \qquad (9.5)$$
$$\text{Anti-selfdual } \Phi = (1 - Z_c^3/6)\phi + Z_c(1 - Z_c/2)\varphi$$
$$+ (\bar{\chi}\zeta\zeta + \bar{\zeta}\bar{\zeta}\chi)(1 - Z_c) + \bar{\zeta}\rho\zeta(1 + Z_c); \qquad (9.6)$$

$$\text{Selfdual } \Psi = (\zeta^{\bar{0}}\nu + \zeta^{\bar{4}}\ell)(1 + Z_c^2/2) + \bar{\zeta}\bar{\zeta}\zeta\chi + \bar{\zeta}\bar{\zeta}\bar{\zeta}\omega + \text{h.c.};$$

$$(9.7)$$

$$\text{Anti-selfdual } \Psi = (\zeta^{\bar{0}}\nu + \zeta^{\bar{4}}\ell)(1 - Z_c^2/2) + (\zeta^{\bar{0}}\nu' + \zeta^{\bar{4}}\ell')Z_c + \text{h.c.}$$

$$(9.8)$$

h.c. signifies hermitian conjugates of property, ω is a QCD singlet state and χ is a chromic sextet.

9.4 Selfdual combinations

So far we have refrained from spelling out the quantum numbers for the various properties but have only implied them obliquely. No more prevarication. In order to merge the duality combinations for QCD SU(3) and electroweak SU(2) we need to identify them clearly and will need to fold in handedness for property mixtures that involve leptons. Let ζ_L and ζ_R refer to left and right leptonic doublet properties respectively. They spawn the singlet scalars $Z_L = \bar{\zeta}_L\zeta_L$ and $Z_R = \bar{\zeta}_R\zeta_R$ and in turn this can produce selfdual combinations like $\zeta_R(1 \pm Z_L)$, etc., effectively doubling the number of states that involve leptonic property.

State counting can become a rather technical exercise in selecting appropriate selfdual combinations. To avoid getting enmeshed with details we will simply refer the reader who is interested to look at Reference [57] and focus on the presently known generations of quarks and leptons. To kick off let us spell out the (obvious) quantum numbers of the first generation $(\zeta^0, \zeta^1, \zeta^2, \zeta^3, \zeta^4)$ which mimic the well-known particles $(\nu_e, d_R, d_G, d_B, e^-)$. Their charges Q are of course $(0, -1/3, -1/3, -1/3, -1)$ while their *conventional* fermion numbers F are $(1, 1/3, 1/3, 1/3, 1)$ although we do not gauge the corresponding operator — in contrast to Q which is effectively gauged through the electroweak sector. Observe that we have ignored the up quarks and the reason is that it will emerge as a mixture of properties. We do indeed come across u-type accompanied by d partners, as I will presently show; but we get more than we bargained for.

To simplify the argument, as before let ζ_L and ζ_R refer to left and right-handed lepton-like property *doublets*. The weak isopsin

$\mathbf{T}_L$ will be gauged in due course, but not its right-handed counterpart, meaning that ζ_R will look like a singlet when viewed through left-handed spectacles. As always the relation between charge Q, isospin T_3 and hypercharge Y is given by $Q = T_3 + Y/2$. Let ζ_c refer to the colour D-type triplet $(\zeta^1, \zeta^2, \zeta^3)$, but weak isodoublets, with $Y = 1/3$, $T_3 = -1/2$ and $Q = -1/3$.

We immediately recognise several incarnations of left chirality leptons as product combinations:

$$\zeta_L \cdot \{1, Z_L, Z_R, Z_R^2\} \cdot \{Z_c, Z_c^2, Z_c^3\},$$

a veritable plethora of states! This is where duality comes into play as we can group the set into just 4 (anti-selfdual) generations:

$$\zeta_L \cdot (1 - Z_L) \cdot \{Z_R, (1 - Z_R^2/2)\} \cdot \{(1 - Z_c^3/6), Z_c(1 - Z_c/2)\}. \tag{9.9}$$

Therefore the prediction is for *another generation of charged leptons and neutrinos*, besides the familiar three of the Standard Model.

Regarding up and down quarks, this attribute scheme also diverges from the Standard Model. The basic building block for a weak quark doublet generation is $\zeta_{L,R}\bar{\zeta}_c\bar{\zeta}_c$ which may be multiplied by various single Z factors to produce selfdual combinations, for instance $(u, d) \sim (\zeta_{L,R}^0 \zeta^{\bar{1}}\zeta^{\bar{2}}, \zeta_{L,R}^4 \zeta^{\bar{1}}\zeta^{\bar{2}})$. But there is another possible basic (u, d) mixture, namely

$$\zeta_c \cdot (\zeta^{\bar{4}}\zeta^0, (\zeta^{\bar{4}}\zeta^4 - \zeta^{\bar{0}}\zeta^0)/\sqrt{2}),$$

leaving a 'loose-end' combination $\mathcal{X} \sim \zeta_c \zeta^{\bar{0}}\zeta^4$. This implies *a new quark* with a charge of $-4/3$ as part of an isotriplet, not doublet. This is a clearcut prediction of combining properties to produce generations. So far there is no evidence for another quark carrying charge $-4/3$ nor for another charged lepton, but there are tantalising indications of another neutrino. Possibly the putative missing states have larger masses: the usual escape route. Time will tell and higher energy machines (especially in the form of lepton–antilepton colliders) may provide some clues via new cross-section thresholds.

Chapter 10

Curving Space–Time–Property

This and the next chapter contain some hairy mathematics, so gird your loins. I will try my best to explain what the various algebraic expressions and equations mean so as not to lose you. For practitioners of GR it will be a walk in the park and I hope that familiarity does not breed contempt.

Einstein geometrised Newtonian gravitation by realising that gravitational forces are really due to the distortion of the space-time continuum by mass/energy and that the magnitude of gravitational acceleration is connected to the curvature of the continuum, as reflected in the metric g_{mn}. The curving of spacetime in 4D is difficult to imagine but it can be ascertained by its geometrical consequences. For example an insect crawling on the surface of a sphere (Earth) can determine the curvature by drawing a triangle with geodesic sides and finding that the sum of the interior angles exceeds 180 degrees. Alternatively the traveller can do two successive moves in different directions in opposite order on the surface and find that one does not end up in the same place. See Figure 10.1; indeed that is one method of working out the (Riemann) curvature (tensor).

The curvature itself can be worked out from the metric g which governs the infinitesimal distance squared between two points separated in spacetime by dx:

$$ds^2 = dx^m dx^n g_{nm} \qquad (10.1)$$

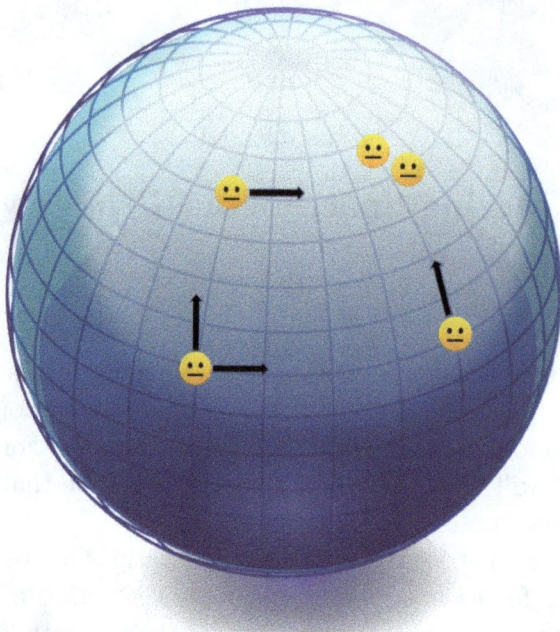

Fig. 10.1 Moving from a point on the equator 500 km east and then another 500 km north and repeating the process in the reverse order leads to different locations. The relative final displacement gives a measure of the Earth's curvature.

with g necessarily symmetric in its indices because the dx are bosonic variables and commute with one another.[1] Textbooks on differential geometry teach us that the 4-index Riemann curvature tensor is given by

$$R_{klm}{}^{j} = \partial_{l}\Gamma_{km}{}^{j} - \partial_{m}\Gamma_{kl}{}^{j} + \Gamma_{km}{}^{n}\Gamma_{nl}{}^{j} - \Gamma_{kl}{}^{n}\Gamma_{nm}{}^{j}, \qquad (10.2)$$

where the Christoffel connection

$$\Gamma_{mn}^{k} \equiv g^{kl}[\partial_{n}g_{lm} + \partial_{m}g_{ln} - \partial_{l}g_{mn}]/2,$$

exhibits its dependence on the metric and derivative ∂g. The connection Γ takes into account the alteration of an orthogonal basis as

[1] Perhaps an example which will be familiar to you is the (distance)^2 between two nearby points on the surface of a sphere of radius ρ, described by a latitude coordinate θ and longitude ϕ when $ds^2 = \rho^2(d\theta^2 + \sin^2\theta\, d\phi^2)$. Here the dx refer to angular changes.

one moves from point to point because of the underlying curvature of space.

The full Riemann tensor in Eq. (10.2) carries four indices and is quite complicated involving double derivatives of the metric. From it one may derive a contracted two-index 'Ricci' curvature, $R_{km} = R_{klm}{}^{l}$ and a fully contracted Ricci scalar curvature $R^{[g]} \equiv g^{mk} R_{km}$. It is the latter quantity which appears in the Einstein–Hilbert Lagrangian for pure gravity.

10.1 Extending geometry to property

It may seem like an odd idea, but we can use the same strategy to describe the curvature in property coordinates. I am going to use smileys, representing attributes, to illustrate the general concept. Suppose we start off with an equable or balanced personality such as pessimistic and optimistic (= antipessimistic) at the same time. Then bad things happen to the neutral state causing anger, succeeded by sadness, resulting an undesirable angry–sad state. An alternative scenario would be to experience sadness followed by anger, also leading to an unhappy sad–angry state. If anticommuting property coordinates are attached to the personality and property space is 'curved' then the two final states may be somewhat different, and the discrepancy would be a signal of property curvature. Figure 10.2 illustrates this.

Much the same thing applies when changing location and altering property by rephasing the attribute. The alteration in trait with x is normally associated with a 'gauge change' and is compensated by the gauge potential A. In fact the curvature of property with spacetime is nothing other than the 'curl' of A and finds its way in the Maxwell tensor F as we shall presently show. Figure 10.3 is a graphic illustration of this argument. We start off with a happy condition, alter its phase and get an altered happy state, in contrast to moving to a new location and rephasing differently there, resulting in a different happy condition.

In Figure 10.4 we have combined the result of curving spacetime: a neutral property is displaced along two different paths, in opposite

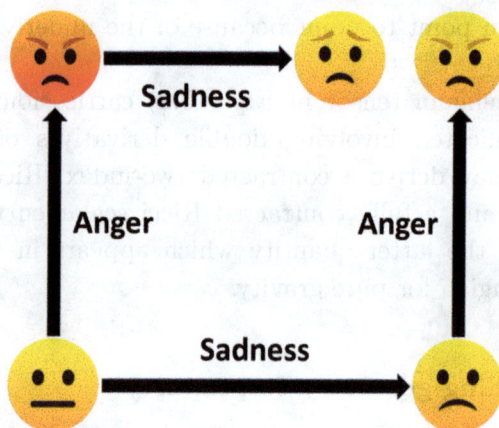

Fig. 10.2 Starting from neutral, moving on to sad and then angry. Contrast with the opposite order: angry followed by sad. The final personalities are different if property space is curved.

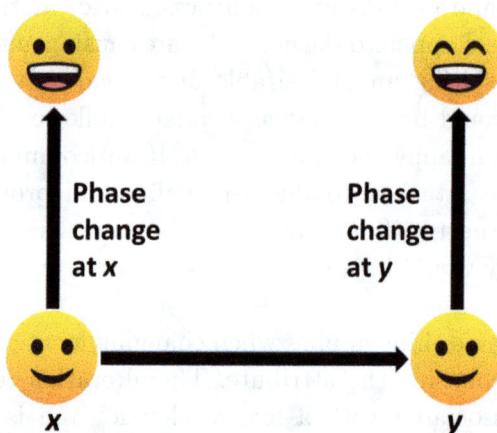

Fig. 10.3 Starting from happy attribute at x, rephasing and getting another happy condition. In contrast, moving to a new position y and rephasing differently, resulting in a new happy atribute.

order and with changed attributes in opposite order. Our job now is to implant some mathematics into such considerations, and more especially to extend the general relativistic formalism to properties, i.e. anticommuting coordinates ζ^μ. Fortunately the pioneers [17, 58] of 'supergravity' found a nice method to do this and we shall mimic

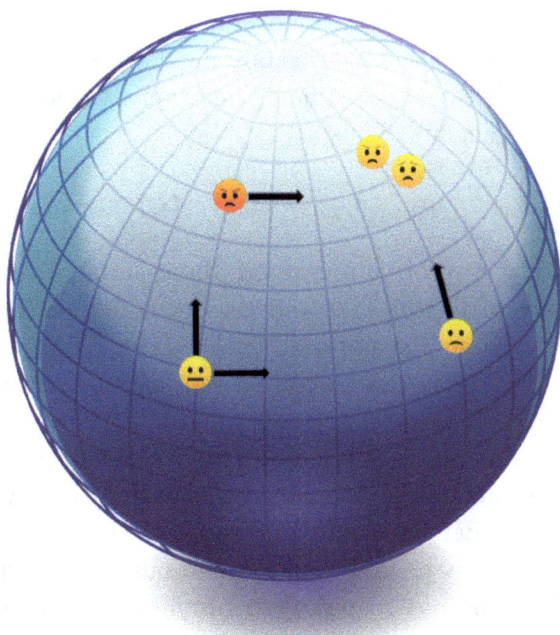

Fig. 10.4 Starting from a neutral disposition at equator, moving east and then north (or *vice versa*) as well as changing the characteristic condition from neutral to sad then angry (and *vice versa*) resulting in new locations and attributes when there is curvature in spacetime and property.

what they concocted in our own particular way, trying to make the procedure less demanding of readers.

10.2 Notational niceties

If we are going to characterise an event fully through 'when–where–what' coordinates, we need to combine them all into a supercoordinate. Upon combination we shall use capital letters in order to avoid any confusion later; also we will use Latin lower case letters for spacetime commuting x and Greek letters for anticommuting property ζ. Thus define the supercoordinate $X^M \equiv (x^m, \zeta^\mu, \zeta^{\bar{\mu}})$ and event squared separation $ds^2 = dx^M dx^N G_{NM}$ or, explicitly,

$$ds^2 = dx^m dx^n G_{nm} + dx^m d\zeta^\nu G_{\nu m} + dx^m d\zeta^{\bar{\nu}} G_{\bar{\nu}m}$$
$$+ d\zeta^\mu d\zeta^\nu G_{\nu\mu} + d\zeta^{\bar{\mu}} d\zeta^\nu G_{\nu\bar{\mu}} + d\zeta^{\bar{\nu}} d\zeta^{\bar{\mu}} G_{\bar{\nu}\bar{\mu}}. \qquad (10.3)$$

Note that $G_{nm}, G_{\nu\mu}$ and $G_{\nu\bar{\mu}}$ are overall bosonic but $G_{\nu m}$ and $G_{\bar{\nu}m}$ are overall fermionic because ds^2 is bosonic. Furthermore the commutation relations between coordinates determine the symmetry properties of the G elements. Thus

$$G_{mn} = G_{nm}, \quad G_{\mu n} = G_{n\mu}, \quad G_{\mu\bar{\nu}} = -G_{\bar{\nu}\mu}. \qquad (10.4)$$

There is a nifty way of summarising these sorts of relations with the appearance of factors of 1 and -1 depending on the nature of the coordinates. Use the notation $[\mu] = 1, [m] = 0$ as aspects of $[M]$, so that Eq. (10.4) can be compacted into the relation $G_{MN} = (-1)^{[M][N]}G_{NM}$.

One can now steam ahead and develop a differential 'graded' tensor calculus. A substantial formalism ensues and those who wish to follow all the gory details should go through the books and published articles [59]. The only thing I wish to point out is that when working out the details, one has to pay *very careful* attention to the order of the indices as they basically affect the $\pm$ factors. For those who do not want to wallow in algebra I simply ask you to take the key formulae below on trust, as they have been thoroughly checked; just skip to Section 10.3 if you find this section too intricate or boring. An important point to make is that our convention is to take all derivatives as left derivatives and not complicate the issue with right derivatives. (This is the standard convention found in most calculus texts.)

The formulae that matter are stated below and I emphasise again that the order of indices as written is very important.

Coordinate changes: When transforming from X to X' for contravariant and covariant vectors respectively:

$$V'^M(X') = V^R(X)\left(\frac{\partial X'^M}{\partial X^R}\right), \quad A'_M(X') = \left(\frac{\partial X^R}{\partial X'^M}\right)A_R(X),$$
$$(10.5)$$

leading to the invariant $V^R(X)A_R(X) = (-1)^{[R]}A_R(X)V^R(X)$.

Metric transformation: The transformation rule below — which also applies to tensors $T_{NM}(X)$ — can be deduced via the direct product of two covariant vectors:

$$G_{NM}(X) = \left(\frac{\partial X'^R}{\partial X^M}\right)\left(\frac{\partial X'^S}{\partial X^N}\right) G'_{SR}(X')(-1)^{[N]([R]+[M])}$$

(10.6)

and it applies to all components of G in one fell swoop.

Inverse metric: This is defined by the rule $G^{LM}G_{MN} = \delta^L{}_N$ with *indices in that particular order*. It satisfies the symmetry property $G^{MN} = (-1)^{[M][N]}G^{NM}$ too; the inverse metric can be used to raise and lower indices according to the rule $V^R \equiv G^{RS}V_S$.

Derivatives: The ordinary derivative is conventionally summarised by the comma notation, $T_{...,M} = (\partial/\partial X^M)T_{...} \equiv \partial_M T_{...}$, whereas the covariant derivative is written $T_{...;M} \equiv D_M T_{...}$ with semicolons and they ensure that the result transforms correctly when the space is curved. We will presently show how D acts on various tensors.

Christoffel connections: For covariant vectors one finds that

$$A_{M;N} \equiv (-1)^{[M][N]}A_{M,N} - A^L\Gamma_{\{MN,L\}} \quad \text{where} \quad (10.7)$$

$$\Gamma_{\{MN,L\}} \equiv \big[(-1)^{([L]+[M])[N]}G_{LM,N}$$
$$+ (-1)^{[M][L]}G_{LN,M} - G_{MN,L}\big]/2. \quad (10.8)$$

Correspondingly for contravariant vectors,

$$A^M{}_{;N} = (-1)^{[M][N]}\left(A^M{}_{,N} + A^L\Gamma_{LN}{}^M\right) \quad \text{where} \quad (10.9)$$

$$\Gamma_{MN}{}^K \equiv (-1)^{[L]([M]+[N])}\Gamma_{\{MN,L\}}G^{LK}$$
$$= (-1)^{[M][N]}\Gamma_{NM}{}^K. \quad (10.10)$$

Riemann curvature: This 4-index tensor $\mathcal{R}$ is obtained through the relation

$$A_{K;L;M} - (-1)^{[L][M]}A_{K;M;L} \equiv (-1)^{[K]([L]+[M])}\mathcal{R}^J{}_{KLM}A_J,$$

(10.11)

which has to do with the difference resulting from a couple of changes done in opposite order. The explicit expression, given in terms of the Christoffel connection, is

$$\mathcal{R}^J{}_{KLM} = (-1)^{[K][M]}(\Gamma_{KM}{}^J)_{,L} +$$
$$(-1)^{[M]([K]+[L])+[K][L]}\Gamma_{KM}{}^N\Gamma_{NL}{}^J$$
$$- (-1)^{[L][M]} \times \text{(above with } L \leftrightarrow M). \quad (10.12)$$

Perhaps a neater way of writing this is via the fully covariant Riemann tensor, $\mathcal{R}_{JKLM} \equiv (-1)^{([J]+[K])[L]}\mathcal{R}^N{}_{KLM}G_{NJ}$ which has the pleasing properties:

$$\mathcal{R}_{JKLM} = -(-1)^{[L][M]}\mathcal{R}_{JKML} = -(-1)^{[J][K]}\mathcal{R}_{KJLM},$$
$$0 = (-1)^{[J][L]}\mathcal{R}_{JKLM} + (-1)^{[J][M]}\mathcal{R}_{JLMK}$$
$$+ (-1)^{[J][K]}\mathcal{R}_{JMKL},$$
$$\mathcal{R}_{JKLM} = (-1)^{([J]+[K])([L]+[M])}\mathcal{R}_{LMJK}.$$

Ricci Curvature tensor: This is simply found through contraction:

$$\mathcal{R}_{KM} \equiv (-1)^{[J]+[K]([L]+[J])+[M]}G^{LJ}\mathcal{R}_{JKLM}$$
$$= (-1)^{[K][M]}\mathcal{R}_{MK}. \quad (10.13)$$

The fully contracted scalar curvature is $\mathcal{R} \equiv G^{MK}\mathcal{R}_{KM}$ and it features in the 'all-events' Lagrangian much like the Hilbert–Einstein Lagrangian does for pure gravity.

We appreciate that the above formalism may be indigestible for those with no appetite for mathematical detail, but I should point out that these formulae are crucial for working out the whole set of possible events and they encapsulate all the forces of nature in compact form once the metric for *where–when–what* is specified.

10.3 Realising Einstein's dream

After general relativity Einstein's primary goal was to unify gravity with electromagnetism and he devoted the last years of his illustrious career towards achieving that marriage. Although Klein and Kaluza has suggested the way forward, Einstein was quite circumspect in following that route because he realised that a full version of that scheme would lead to a scalar–tensor theory of gravity [60], which he disfavoured. Besides, he did not approve of the infinity of additional states which result from excitations due to extra hidden degrees of freedom pertaining to new bosonic dimensions.

Below I will describe our own attempts[2] to realise this particular unification dream and which then naturally leads to the marriage of gravity with all other *observed* natural forces. Where we depart from other researchers is that we attach extra *anticommuting scalar* coordinates to spacetime that represent the characteristics of events and any exchanged properties. It would have been more natural to attach spin 1/2 degrees of freedom, as per conventional supersymmetry (SUSY) as in supergravity, but the trouble is that having a set of independent θ_α^n would then lead to problematic spin proliferation. In any case SUSY has fallen from grace as there has been no experimental evidence in its favour for over 40 years, despite its numerous adherents.

In this section I will show how adding just one extra fermionic property, namely electricity, achieves Einstein's desired marriage of gravity with electromagnetism. It is ironic to think that if Kaluza–Klein and Einstein had taken the extra fifth coordinate to be complex fermionic the dream would have been achieved. Thus we shall add a single ζ coordinate (and of course its conjugate $\bar\zeta$) representing electricity, to spacetime x, as explained in Section 10.2. Every result that ensues hinges on the form of the generalised metric G_{MN} so we need to spell its features in general terms.

Since we are only dealing with one property, we will ignore the indices μ that characterise several attributes; after all there is only one attribute for electricity: $\mu = 4$. The event separation squared can thereby be written as

$$ds^2 = dx^m dx^n G_{nm} + dx^m d\zeta\, G_{\zeta m} + dx^m d\bar\zeta\, G_{\bar\zeta m} + d\bar\zeta d\zeta\, G_{\zeta\bar\zeta}.$$

---

[2]Most of the material below was carried out in collaboration with Paul Stack, who devised an algebraic computer program to carry out the required complicated calculations.

If the universe were static and thus uneventful we would encounter a flat space with

$$ds^2 = dx^m dx^n \eta_{nm} + \ell^2 d\bar{\zeta} d\zeta,$$

wherein we have rustled in a length scale ℓ to make sure that the event separation has uniform length dimensions, based upon the assumption that property is a dimensionless concept. Therefore $G_{\zeta\bar{\zeta}} = -G_{\bar{\zeta}\zeta} = \ell^2$. Readers will note that the metric is invariant under global phase transformations, $\zeta \to \zeta' = \exp[i\theta]\zeta$ that do not mix spacetime with property. However once we make the local changes, $\zeta \to \zeta' = \exp[i\theta(x)]\zeta$, we come across a contradiction in as much as the transformation rule (10.6) is violated. (This is not entirely surprising because we are intrinsically mixing location and property.) There is one way to fix this deficiency and it is to pay attention to the other elements of the metric and particularly the elements $G_{\zeta m}$ and $G_{\bar{\zeta}m}$, as we will do in Section 10.3.2.

10.3.1 *Metric elements and frame vectors*

One needs to ask: what is the form of the off-diagonal element $G_{\zeta m}$? It has to be vectorial (via index m) and must be overall fermionic. There is basically only one possible choice, namely a product of a vector field A_m and a property coordinate; for reasons which will be clear quite soon, we take

$$G_{m\zeta} = -ie\ell^2 \bar{\zeta} A_m / 2, \quad G_{m\bar{\zeta}} = -ie\ell^2 A_m \zeta / 2. \tag{10.14}$$

The scale ℓ and the electron charge e enter because (i) the electromagnetic potential is *always* accompanied by that charge; also (ii) the scale ℓ, ultimately tied to Newton's gravitational constant G_N, makes an appearance to ensure the correct mass dimensions. This has repercussions in the G_{mn} sector, which can be appreciated by writing the metric[3] in terms of orthogonal frame vectors $\mathcal{E}$ or the inverse superbeins E. [These frame vectors are helpful when one

---

[3]Early letters of the alphabet connote flat space so I corresponds to the flat metric as stated earlier.

comes to constructing the Lagrangians for fermions.] The metric can be regarded as a contraction over frame vectors $\mathcal{E}$:

$$G_{MN} \equiv (-1)^{[A][N]} \mathcal{E}_M{}^A \mathcal{E}_N{}^B \mathcal{I}_{BA} \qquad (10.15)$$

$$G^{MN} = (-1)^{[B][M]} \mathcal{I}^{BA} E_A{}^M E_B{}^N \qquad (10.16)$$

$$\mathcal{E}_M{}^B E_B{}^N = \delta_M{}^N, \qquad (10.17)$$

with

$$\mathcal{E}_m{}^a = e_m{}^a, \quad \mathcal{E}_m{}^\zeta = -ieA_m\zeta, \quad \mathcal{E}_m{}^{\bar\zeta} = ie\bar\zeta A_m, \qquad (10.18)$$

and diagonal unity in the ζ sector of $\mathcal{E}$. In Eqs. (10.15) and (10.16) we recognise that the frame vectors and vielbeins are in a sense the square roots of the metric tensor summed over locally flat Lorentz indices. They enter prominently when one handles the Dirac spinor in curved space when the gamma matrices in flat space γ_a that obey the flat space Clifford algebra $\{\gamma_a, \gamma_b\} = \eta_{ab}$ generalise to $\gamma_m \equiv e_m{}^a \gamma_a$ with $\{\gamma_m, \gamma_n\} = g_{mn}$ in curved space. Moreover they induce a contribution to the full spacetime metric:

$$G_{mn} = \mathcal{E}_m{}^a \mathcal{E}_n{}^b \eta_{ba} + \mathcal{E}_m{}^\zeta \mathcal{E}_n{}^{\bar\zeta} \ell^2 = g_{mn} + e^2 \ell^2 \bar\zeta A_m A_n \zeta$$

10.3.2 Gauge transformations

We mentioned before that the metric transformation rule goes awry under local gauge changes. This is where the x–ζ elements of the metric (10.14) come to the rescue. If we focus on the element $G_{m\zeta}$, or equivalently $G_{m\bar\zeta}$, with the identification (10.14) the transformation rule (10.6) corresponds no more and no less to the gauge transformation $A'_m = A_m + \partial_m\theta$ under a local phasing by θ of ζ. This makes very good sense: the space–time–property metric element (embodied in the vector gauge field), serves as a messenger, transporting property from one location to another. The only freedom left to us in specifying the G_{MN} elements is the ability to multiply terms by the scalar invariant $Z = \bar\zeta\zeta$ since it is not affected by phasing. We can therefore multiply G_{mn} by a factor $(1 + cZ)$ and $G_{\bar\zeta\zeta}$ by $(1 + c'Z)$, where c, c' are constants, which can be viewed as expectation values of background scalar fields ϕ, ϕ'; they are similar to those arising in Kaluza–Klein (KK) theory in the fifth dimension. The advantage which this scheme has over KK is that there are no infinite modes to contend with. This restricted freedom finds its way into the value of the cosmological constant, as we will see.

10.3.3 *The full Lagrangian*

The full action is constructed, just as in general relativity, through a simple integration over spacetime and property of the full Ricci scalar curvature $\mathcal{R}$. Thus the Lagrangian density is obtained from the attribute integral (up to an overall scaling constant):

$$\mathcal{L} \propto \int d\zeta d\bar{\zeta} \sqrt{-G..} \, \mathcal{R}; \quad \mathcal{R} = G^{MK}\mathcal{R}_{KM} = G^{MK}(-1)^{[K][L]}\mathcal{R}^{L}{}_{KLM},$$

where $\sqrt{-G..}$ is the Berezin superdeterminant[4] of the full metric. Evaluating $\mathcal{L}$ involves meticulous calculation. With any number of properties it always reduces to the form

$$\mathcal{L} = \sqrt{-g..}[\mathcal{A} \, R^{[g]} + \mathcal{B} \, \mathrm{Tr}(F_{mn}F^{mn}) + \mathcal{C}], \quad (10.19)$$

namely the sum of the ordinary gravitational curvature, the Maxwell Lagrangian for electromagnetism, plus a cosmological contribution. The coefficients of the terms for a single ζ are

$$\mathcal{A} = 2(c - c')/\ell^2, \quad \mathcal{B} = -e^2/2, \quad \mathcal{C} = 24(c - c')^2/\ell^2.$$

This needs to be compared with the textbook Lagrangian,

$$\mathcal{L} = \sqrt{g..} \, [R^{[g]} - \mathrm{Tr}\,(F_{mn}F^{mn})/4 - 2\Lambda]/16\pi G_N. \quad (10.20)$$

Therefore if we scale (10.19) by a factor of $1/2e^2$ to get the correct normalisation for electromagnetism we can reproduce the standard result with

$$G_N = \frac{e^2\ell^2}{16\pi(c - c')}, \quad \Lambda = -\frac{96\pi G_N(c - c')^2}{e^2\ell^2}.$$

Because G_N must be positive we need $c > c'$, but this does not help with the value of the cosmological constant which is experimentally

---

[4]See the glossary for a description of how this is to be calculated.

found to be very small and positive. This is just a minor blemish on our scheme but we can live with it because we have neglected the impact of the other forces of nature with their own properties. Those bring in further coefficients c multiplying the various Z scalars and it becomes simple to reverse the sign of the cosmological term as the next chapter will demonstrate. In any event we get basically the same format as KK theory but *without infinite modes*. The gravitational constant is intimately connected with the scale ℓ connecting property with spacetime. In the absence of matter we can thereby reproduce the standard equations of motion for gravity and the electromagnetic field, which was our goal.

Chapter 11

Geometry of Events

This section is concerned with extending the geometrical picture of electromagnetism just described to all the other known forces: the denouement of an eventful journey. We will start with QCD, marry it with electromagnetism before tackling the electroweak standard picture with its peculiar handedness. Finally we shall be combining them all into one full-scale 'where–when–what' package. So hold on to your horses as the route brings in some bumpy mathematics.

11.1 Chromic geometry

It took a while before the QCD fundamental picture of strong interactions was established. First off, three separate colours were invoked to comprehend how the quarks in baryon and meson ground states could hold together without violating the spin-statistics theorem. Next, the results from high energy lepton–hadron scattering clarified how quarks were trapped into white or colourless blends via QCD gluonic forces. Lastly the physical evidence for gluons and colours came from the discovery of multiparticle production in the form of two or three narrow jets. But the most pleasing aspect of the accepted theory was that it was a gauge model, like gravity and electromagnetism and could be handled in the same way; it involved strong couplings which get weaker as the distances decrease and it means that strong processes can be treated perturbatively at high energy. However QCD brings in certain complications in as much as it is a non-Abelian theory — meaning that the chromic transformations do not commute. This aspect was somewhat unfamiliar to theorists at

that time but now physicists are more comfortable with its ramifications and handle it with relative ease.

To understand a little better how these QCD forces operate
between quarks, we will label the chromic properties $(\zeta^1, \zeta^2, \zeta^3)$
where: red = 1, green = 2, blue = 3. The QCD phasings and mixings
between ζ^μ can be summarised by the rule

$$\zeta^\mu \rightarrow U^{\mu\bar\nu}\zeta^\nu = (\exp[i\Theta])^{\mu\bar\nu}\zeta^\nu \tag{11.1}$$

where the matrix U is unitary and has unit determinant; mathematically we express this as U(3) = SU(3)⊗U(1) with $UU^\dagger = 1$ and
$\det(U) = 1$ for SU(3). A crude way of imagining what Eq. (11.1) does
is to think of a red–green–blue palette; the U(1) transformations correspond to phasing in the same way the amalgam of all three colours,
whereas the SU(3) transformations can be vaguely understood as a
proportionate mixing of the three colours separately, with the pro-
portions summing to unity. Thus U(1) changes are roughly like dark-
ening or lightening the overall colour whereas SU(3) transformations
correspond to changing the hue.

We really have to introduce the three chromic labels 1,2,3 to make
more progress. Observe that $Z_c = \zeta^{\bar\mu}\zeta^\mu$ is invariant under the rotations (11.1) because U is unitary and therefore we can construct three
independent invariant polynomials $Z_c, Z_c{}^2, Z_c{}^3$. Higher powers of Z_c
vanish identically from the Grassmannian nature of property. This
becomes important when we come to consider the chromic metric
associated with colour and spacetime separation:

$$ds^2 = dx^m dx^n G_{nm} + dx^m d\zeta^\nu G_{\nu m} + dx^m d\zeta^{\bar\nu} G_{\bar\nu m}$$

$$+ d\zeta^\mu dx^n G_{n\mu} + d\zeta^\mu d\zeta^\nu G_{\nu\mu} + d\zeta^\mu d\zeta^{\bar\nu} G_{\bar\nu\mu}$$

$$+ d\zeta^{\bar\mu} dx^n G_{n\bar\mu} + d\zeta^{\bar\mu} d\zeta^\nu G_{\nu\bar\mu} + d\zeta^{\bar\mu} d\zeta^{\bar\nu} G_{\bar\nu\bar\mu}. \tag{11.2}$$

To get the correct non-Abelian transformations for local chromic
changes (11.1) we must, as with electromagnetism, introduce the
hermitian vector gluon gauge field $V^{\mu\bar\nu}$ matrix (with its associated
coupling constant f) in the (ζ, x) sector:

$$G_{m\nu} = -i\ell^2 f(\bar\zeta V_m)^{\bar\nu}/2, \quad G_{m\bar\nu} = -i\ell^2 f(V_m \zeta)^\nu/2, \tag{11.3}$$

which also flows into the spacetime sector:

$$G_{mn} = g_{mn} + f^2 \ell^2 \bar{\zeta} (V_m V_n + V_n V_m) \zeta / 2. \tag{11.4}$$

The last freedom left to us is possible multiplication of the full metric simply by the overall factor $C = 1 + c_1 Z_c + c_2 Z_c^2 + c_3 Z_c^3$. Such a factor is gauge invariant but will impact on the coefficients of the various terms in (10.19). Skipping details of calculation which can be accessed elsewhere [57], we finally arrive at the following coefficients:

$$\mathcal{A} = (96/\ell^6)(2c_1^3 - 3c_1 c_2 + c_3),$$

$$\mathcal{B} = (4f^2/\ell^4)(3c_1^2 - 2c_2),$$

$$\mathcal{C} = -(1152/\ell^8)(5c_1^4 - 10c_1^2 + 2c_2^2 + 3c_1 c_3).$$

It is clear that the Lagrangian can be rescaled to the standard gravitational one plus that of the gluons; also the c_i can be adjusted to yield a cosmological constant that accords with experiment. I should also point out that one automatically gets the non-Abelian version of the Maxwell tensor in addition to the electromagnetic one:

$$F_{mn} = \partial_m V_n - \partial_n V_m - if[V_m, V_n] \tag{11.5}$$

and the non-Abelian equations of motion that result. The non-Abelian version of the stress tensor, with the correct sign, also enters the generalised Einstein tensor automatically:

$$\mathcal{R}_{KL} - G_{KL}\mathcal{R}/2. \tag{11.6}$$

It is a simple process to include QED with QCD. All one needs to do is to include the electrical ζ^4 coordinate with its gauging. This add the further elements $G_{m4} = -i\ell^2 e\zeta^{\bar{4}} A_m/2$, $G_{m\bar{4}} = -i\ell^2 e A_m \zeta^4/2$ and $\ell^2 e^2 \zeta^{\bar{4}} A_m A_n \zeta^4$ to (11.4). The latter are themselves generated from the frame vectors

$$\mathcal{E}_m{}^\kappa = -i(fV_m - eA_m/3)^{\kappa\bar{\iota}}\zeta^\iota, \quad \mathcal{E}_m{}^4 = ieA_m \zeta^4$$

that are appropriate for down quarks plus electrons. Notice that we have *excluded* gauge fields which bridge colour labels 1,2,3 with electrical label 4; the point being that nature has picked the gauge group to be the direct product group SU(3)⊗U(1) *not* SU(4) as far as we can tell. Were we to relax this condition we would be in the realm of grand unified theory with all its apparent paradoxes.

11.2 Handling handedness

It must have come as a great shock when beta decay was discovered to preferentially favour the left hand; before then it was *assumed* that all the forces could not distinguish between left and right chirality. While it is true that in certain organic compounds or crystals one or other chirality is prominent, it is argued that this is primarily an accident of initial conditions which assisted the development of one chirality rather than the other and is not because the forces in question break chiral symmetry. Early on, with the realisation that parity is indeed broken in weak decays, scientists could console themselves with the fallback position that *in combination with charge conjugation* C the so-called CP symmetry is preserved. However that posture did not last long when it was found that neutral K-mesons (and more recently B^0-meson decays) show no inclination to preserve CP in their decay modes. So far the only symmetry that seems to hold good is CPT symmetry; it has been checked via delicate experiments on the properties of antiparticles and is an automatic consequence of a local Lagrangian which respects special relativity.

Thus we have had to come to terms with the violation of certain discrete symmetries and in particular the need to distinguish left from right in a basic way. In this section I will try to implement this in the framework of the event scheme. Neutrinos from beta decay always come out left-handed, though there surely exist right-handed components: we know this to be true because neutrinos are not exactly massless and undergo flavour transmutations as they travel. Previously I introduced the doublet of electricity plus neutrinicity in the pair of properties (ζ^0, ζ^4). Weak interactions oblige us to distinguish their characteristic handedness, via *two* independent doublets ζ_L, ζ_R of which only the former features in beta decay. Readers will argue: what about the hadrons and especially the quarks?

Well, in our scheme the quarks such as $u_{\text{blue}} \sim \zeta^0 \zeta^{\bar{1}} \zeta^{\bar{2}}$ are built up of chromicity property in combination with leptonicity so we can carry over the idea to the strongly interacting particles too without a great deal of trouble; so for instance, $d_{L\text{blue}} \sim \zeta_L^4 \zeta^{\bar{1}} \zeta^{\bar{2}}$. More tricky is the third generation, when we come across a combination of two leptonic properties. If we are considering an up quark we will need to take both neutrinicity and electricity components to be left; otherwise parity schizophrenia would ensue. I must confess that it would have been simpler if the property coordinates themselves carried spin 1/2 labels for then chirality would be a natural attribute, but a fall into a spin morass would also have been the price to pay.

Setting aside the gauge group, the quantum numbers that accompany the various properties have to be as shown in the table to conform with the Standard Model assignments:

Property	Q	T_{3L}	Y
ζ_L^0	0	1/2	−1
ζ_L^4	−1	−1/2	−1
ζ_R^0	0	0	0
ζ_R^4	−1	0	−2
$\zeta^{1,2,3}$	−1/3	−1/2	1/3

In the above table Q signifies charge, Y means weak hypercharge and T_{3L} is the third component of weak isospin. The conjugate properties (with a bar above the label) are to be associated with antiparticles possessing opposite quantum numbers of course. Such values are simply added in a blended state like the up quark in the third generation: $\zeta^1 \zeta^{\bar{4}} \zeta^0$. Although T_{3R} possesses similar quantum numbers to T_{3L} it is not gauged and we have therefore ignored right-handed isospin altogether. If in future it is discovered that there are gauge bosons with right-handed interactions, one can easily accomodate them by extending the table appropriately. Given separate chiral properties we can envisage a combined property polynomial which is invariant under either left or right rotations:-

$$C = C_R C_L = [1 + c_R Z_R + c_{RR} Z_R{}^2][1 + c_L Z_L + c_{LL} Z_L{}^2]. \quad (11.7)$$

We will shortly find out where this leads to when we study electroweak geometry.

11.3   Electroweak curvature

With left and right fields separated one can impose SU(2) duality symmetry on each chirality. The fermionic superfield thereby admits the property expansion

$$2\Psi = \bar{\zeta}_L[\psi_L(1 + Z_R^2/2) + \psi'_L Z_R](1 + Z_L) + (L \leftrightarrow R) \quad (11.8)$$

$$2\bar{\Psi} = [\overline{\psi_L}(1 + Z_R^2/2) + \overline{\psi'_L} Z_R]\zeta_L(1 + Z_L) + (L \leftrightarrow R), \quad (11.9)$$

which introduces two lepton generations — but remember we have not considered the effect of chromicity at this stage which can augment the number of generations. The chirality projections $(1 \pm i\gamma_5)/2$ ensure that no mass terms can arise from the product $\bar{\Psi}\Psi$ because there are not enough powers of property to give a nonzero answer when integrating over the several ζ; this is a bonus as it stipulates that we need to couple Ψ to a boson superfield Φ in order to generate a mass term. The kinetic term emerges correctly however; in flat spacetime one finds

$$-\int d^2\zeta_R \ldots d^2\bar{\zeta}_L \ \bar{\Psi} i\gamma \cdot \partial\Psi = \overline{\psi_L} i\gamma \cdot \partial\psi_L + \overline{\psi'_L} i\gamma \cdot \partial\psi'_L + (L \leftrightarrow R).$$

As far as the bosons are concerned there is an analogous expansion to (10.7). I shall avoid overwhelming the reader with excessive mathematics by just pointing out that it contains a weak isopin quartet (singlet + triplet):

$$2\Phi = [\bar{\zeta}_R \phi \zeta_L + \bar{\zeta}_L \phi^\dagger \zeta_R](1 + Z_L)(1 + Z_R) + \cdots \quad (11.10)$$

because that component contains the Higgs field. Specifically we encounter a matrix $\phi^{\mu\bar{\nu}} = (\phi_0 I + \phi \cdot \tau)^{\mu\bar{\nu}}/\sqrt{2}$ for whom the quantum numbers $T_{3L}, Y, Q = T_{3L} + Y/2$ read

$$2T_{3L}(\phi^{0\bar{0}}, \phi^{0\bar{4}}, \phi^{4\bar{0}}, \phi^{4\bar{4}}) = (-1, 1, -1, 1),$$
$$Y(\phi^{0\bar{0}}, \phi^{0\bar{4}}, \phi^{4\bar{0}}, \phi^{4\bar{4}}) = (1, 1, -1, -1),$$
$$Q(\phi^{0\bar{0}}, \phi^{0\bar{4}}, \phi^{4\bar{0}}, \phi^{4\bar{4}}) = (0, 1, -1, 0). \tag{11.11}$$

This is a good place to introduce the gauge fields of SU(2)⊗U(1) to verify that we can indeed reproduce the results of the Standard Model for leptons (albeit with two generations). Acting on the leptonic doublet coordinates ζ^0, ζ^4 are the combination of gauge fields, $V_m = L_m + R_m$ restricted to

$$L_m = (gW_m \cdot \tau - g'B_m)/2, \quad R_m = g'B_m(\tau_3 - 1)/2. \tag{11.12}$$

g is the weak SU(2) coupling constant attached to the gauge field W and g' is the hypercharge coupling attached to the corresponding field B. These feed into the metric components connecting property to spacetime:

$$G_{m\zeta_L} = -i\ell^2 L_m C/2, \quad G_{m\zeta_R} = -i\ell^2 R_m C/2,$$
$$G_{\zeta_L \bar{\zeta}_L} = G_{\zeta_R \bar{\zeta}_R} = \ell^2 C/2, \, G_{\zeta_l \zeta_R} = G_{\bar{\zeta}_L \bar{\zeta}_R} = G_{\zeta_L \bar{\zeta}_R} = G_{\zeta_R \bar{\zeta}_L} = 0 \tag{11.13}$$

plus the remaining element $G_{mn} = C[g_{mn} + (\text{gauge field terms})]$. This is what we should expect and indeed what transpires.

When working out the full Lagrangian $\mathcal{L}$ there will of course arise the spacetime curvature $R^{[g]}$ plus a cosmological term coming from the property curvature plus a series of gauge field terms arising from $G_{m\zeta}$ contributions to the curvature scalar. These gauge field contributions bring in their associated squared coupling constants. By integrating over left and right coordinates, and noting that the superdeterminant reduces to $\sqrt{-G_{..}} = (2/\ell^2)^4 \sqrt{-g_{..}} (C_R C_L)^{-3}$, one can arrive at [57] the sum of the anticipated three terms:

$$\int d^2 \zeta_R \ldots d^2 \bar{\zeta}_L \sqrt{-G_{..}} \mathcal{R}$$
$$= 36\sqrt{-g_{..}} (2/\ell^2)^4 R^{[g]} (2c_R^2 - c_{RR})(2c_L^2 - c_{LL})$$

$$-\frac{3}{2}\sqrt{-g_{..}}\left(\frac{2}{\ell^2}\right)^3 [c_L(3c_R^2 - 2c_{RR}(g^2\mathbf{W}_{mn}\cdot\mathbf{W}^{mn}$$

$$+ g'^2 B_{mn}B^{mn}) + g'^2 2c_R(3c_L^2 - 2c_{LL})B_{mn}B^{mn}]$$

$$+ 12\sqrt{g_{..}}(2/\ell^2)^5[(4c_{LL}-5c_L^2)(2c_R^2-c_{RR})+(L\leftrightarrow R)],$$

where $W_{mn} = W_{n,m} - W_{m,n} + ig[W_n, W_m]$ and $B_{mn} = B_{n,m} - B_{m,n}$.

It is very important to have a uniform normalisation for the gauge fields above and at the same time ensure that gravity couples universally to them. This requires that

$$c_L(3c_R^2 - 2c_{RR})(g^2 - g'^2) = 2c_R(3c_L^2 - 2c_{LL})g'^2 \qquad (11.14)$$

be obeyed. Moreover if property curvature in the property sector is parity conserving so that $c_R = c_L = c$, $c_{RR} = c_{LL} = c_2$, this will signify that all parity violation arises only from the gauge sector. Hence we obtain a connection between the hypercharge and weak boson couplings: $g^2 = 3g'^2$. The prediction coming from the electroweak sector alone is therefore for a weak mixing angle of 30°. (This will undergo a minor revision when we include the strong sector with its gluon fields.) The proposal to equalise left and right coefficients in C is quite reasonable because the property curvature polynomial accompanies the gravitational field too and, as far as we know, unquantised gravity does not discriminate between the two chiralities.

It only remains to consider the interaction between the gauge fields with matter, specifically the lepton or quark doublets as in (11.7)–(11.10). For that we return to the vielbeins E that generate the metric elements (11.12); these read:

$$\begin{pmatrix} E_a{}^m & E_a{}^\mu & E_a{}^{\bar\mu} \\ E_\alpha{}^m & E_\alpha{}^\mu & E_\alpha{}^{\bar\mu} \\ E_{\bar\alpha}{}^m & E_{\bar\alpha}{}^\mu & E_{\bar\alpha}{}^{\bar\mu} \end{pmatrix}$$

$$= \frac{1}{\sqrt{C}}\begin{pmatrix} e_a{}^m & i[(L_a\zeta_L)+(R_a\zeta_R)]^\mu & -i[(\bar\zeta_L L_a)^{\bar\mu}+\bar\zeta_R R_a)]^{\bar\mu} \\ 0 & \delta_\alpha^\mu & 0 \\ 0 & 0 & \delta_{\bar\alpha}^{\bar\mu} \end{pmatrix}.$$

$$(11.15)$$

The fermion kinetic energy can be written in the form $\bar{\Psi} i \Gamma^A D_A \Psi$ where

$$D_A = E_A{}^M = E_A{}^m \partial_m + E_A{}^\mu \partial_\mu + E_A{}^{\bar\mu} \partial_{\bar\mu}$$

and the consequence of integrating over left and right properties is to yield the expected result for the two leptonic generations:

$$\overline{\psi_L} \gamma^a (i\partial_a + L_a)\psi_L + \overline{\psi_R} \gamma^a (i\partial_a + R_a)\psi_R + (\psi \to \psi').$$

Recalling that $(\psi^0, \psi^4) = (\nu, l)$, one ends up with

$$\mathcal{L}_\psi = \bar{l}\gamma \cdot (i\partial_a - eA)l$$
$$+ \bar{\nu} i \gamma \cdot \partial \nu + \frac{e}{\sqrt{2}\sin\theta}[\overline{\nu_L}\gamma \cdot W^+ l_L + \overline{l_L}\gamma \cdot W^- \nu_L]$$
$$+ \frac{e}{\sin 2\theta}(\overline{\nu_L}\gamma \cdot Z\nu_L) + e\tan\theta(\overline{l_R}\gamma \cdot Z\nu_R) - e\cot\theta(\overline{l_L}\gamma \cdot Z\nu_L)$$
$$+ (l, \nu) \to (l,'\nu'), \tag{11.16}$$

where θ is the weak mixing angle between hypercharge B and isotopic W^0 which in turn produce the A and Z fields. We remind readers that this weak angle is given by

$$\cos\theta = \frac{g}{\sqrt{g^2 + g'^2}}, \quad \sin\theta = \frac{g'}{\sqrt{g^2 + g'^2}}, \quad e = \frac{gg'}{\sqrt{g^2 + g'^2}}.$$

Equation (11.16) simplifies quite a lot when the weak angle equals 30° as demanded by gravitational universality, because the Z field interacts axially with the charged leptons, while the electromagnetic field remains purely vectorial of course.

In order to complete the verification that our scheme accords with the standard description, we must consider the role of the Higgs field as it determines the mass ratio between W and Z vector bosons as well. Therefore consider the action of the covariant derivative on the bosonic superfield Φ:

$$D_a \Phi = [\partial_a + i(V_a \zeta)^\mu \partial_\mu - i(\bar\zeta V_a)^{\bar\mu} \partial_{\bar\mu}]\Phi. \tag{11.17}$$

Identifying Φ as per (11.9), we arrive at

$$2D\Phi \cdot D\Phi = (1 + 2Z_L)(1 + 2Z_R)[\bar{\zeta}_R\{\partial\phi + i(\phi L - R\phi)\}\zeta_L\bar{\zeta}_L$$
$$\times \{\partial\phi + i(\phi R - L\phi)\}\zeta_R], \qquad (11.18)$$

plus terms which disappear when property is integrated. The following expectation values must be taken for the quartet ϕ:

$$\langle\phi_-\rangle = 0, \ \langle\phi_+\rangle = v; \quad \phi_\pm \equiv \phi_0 \pm \phi_3$$

corresponding to $\langle\phi_0\rangle = \langle\phi_3\rangle$. Calculation then shows that

$$\langle 2\text{Tr}[(\phi R - L\phi)(\phi L - R\phi)]\rangle$$
$$\rightarrow v^2 g^2 W^+ W^- /2 + v^2(g^2 + g'^2)Z^2/4$$
$$= \frac{e^2 v^2}{2\sin^2\theta}W^+ W^- + \frac{e^2 v^2}{\sin^2 2\theta}Z^2. \qquad (11.19)$$

Perfect! The electromagnetic field remains massless and the ratio of W mass to Z mass is given by the weak mixing angle θ which is predicted to be $30°$ solely from the electroweak sector.

11.4   Knitting all the strands

We now need to fold in the strong sector and its contributions before finally incorporating spacetime curvature. Including the gluon fields V_m means multiplying (11.7) by the strong curvature polynomial arising in the property sector; namely

$$C' = [1 + c_R Z_R + c_{RR} Z_R{}^2][1 + c_L Z_L + c_{LL} Z_L{}^2]$$
$$\times (1 + c_1 Z_c + c_2 Z_c^2 + c_3 Z_c^3),$$

where $Z_c \equiv \zeta^{\bar{1}}\zeta^1 + \zeta^{\bar{2}}\zeta^2 + \zeta^{\bar{3}}\zeta^3$ is the chromic invariant. Incorporating the gluon V_m in the metric with the electroweak bosons, one obtains

(x, x) sector : $G_{mn} = g_{mn}C + \ell^2\bar{\zeta}(A_m A_n + A_n A_m)\zeta C'/2,$

$$(11.20)$$

(x, ζ_L) sector : $G_{m\zeta_L} = -i\ell^2\bar{\zeta}_L(gW_m \cdot \tau - g'B_m)C'/4,$ (11.21)

(x, ζ_R) sector : $G_{m\zeta_R} = -i\ell^2 \bar{\zeta}_R h' B_m (\tau_3 - 1) C'/4.$ (11.22)

(x, ζ) sector : $G_{m\zeta} = -i\ell^2 \bar{\zeta} \left(f V_m \cdot \lambda - \frac{2}{3} g' B_m \right) C'/4,$ (11.23)

$(\zeta, \bar{\zeta})$ sector : $G_{\mu\bar{\nu}} = \ell^2 \delta_\mu{}^\nu C'/2.$ (11.24)

[We have taken the property polynomial C multiplying the gravitational part to be independent of the polynomial C' characterising the property sector. Later on we will simplify the argument by identifying these polynomials to be equal: $C = C'$.]

Last, we have to integrate over all these properties to arrive at the final Lagrangian. Omitting calculational details, we end up with the gauge field contributions:

$$
\begin{aligned}
\mathcal{B}\mathrm{Tr}(F \cdot F) &= \frac{1}{12}(3c_1^2 - 2c_2)(c_{LL} - c_L^2)(c_{RR} - c_R^2) \\
&\quad \times [f^2 V^{mn} \cdot V_{mn} + 2g'^2 B^{mn} B_{mn}/3] \\
&\quad - \frac{1}{2}(2c_1^3 - 3c_1 c_2 + c_3)c_R(c_{LL} - c_L^2)g'^2 B^{mn} B_{mn} \\
&\quad - \frac{1}{4}(2c_1^3 - 3c_1 c_2 + c_3)c_L(c_{RR} - c_R^2) \\
&\quad \times [g^2 W^{mn} W_{mn} + g'^2 B^{mn} B_{mn}].
\end{aligned}
$$
(11.25)

As before and for the same reasons, set $c_R = c_L$, etc. In order to get a uniform normalisation (and by implication a universal gravitational coupling) we must impose the following relation between strong and weak charged boson couplings: $f^2(2c_2 - 3c_1^2)(c_{LL} - c_L^2) = 3g^2 c_L (2c_1^3 - 3c_1 c_2 + c_3)$. Doing the same for the weak hypercharge field we deduce that the three coupling constants are related by

$$
g^2 = 3g'^2 + 2g^2 g'^2/3f^2.
$$
(11.26)

Alternatively, expressing the weak mixing angle θ via $g' = g \tan \theta, e = g \sin \theta$, we can compact (11.26) to

$$
4\sin^2 \theta = 1 - 2e^2/3f^2 = 1 - 2\alpha_{\mathrm{em}}/3\alpha_{\mathrm{strong}},
$$

where α signify fine structure constants. The consequence is that the leptonic result of $\sin^2 \theta = 1/4$ is diminished. In fact the ratio

$\alpha_{em}/\alpha_{strong}$ runs with energy in logarithmic fashion according to the renormalisation group and so does our estimate. At about 100 GeV we know that $\alpha_{em} \simeq 0.008$ and $\alpha_{strong} \simeq 0.115$, so the prediction from our anticommuting property scheme is $\sin^2 \theta \simeq 0.238$ and this is quite near the experimental estimates, which hover around 0.232. This is without even taking account of quantum corrections.

The impact of the property polynomial C on the quarks and leptons is to induce a mixing of generations but has no effect on the gauge interactions which act uniformly as expected on each of the generations even after we impose duality. This mixing can be redefined to act on the down quarks to accord with the standard description. All that the curvature in the spacetime component g_{mn} does is to yield a Lagrangian that is generally covariant under coordinate transformations. As for the coefficients of the cosmological constant Λ and the spacetime Ricci scalar $R^{[g]}$, there are enough c_i to fix the Newtonian constant in terms of the scale ℓ and to arrange that Λ becomes very small. It must be admitted that the value of Λ is unnaturally miniscule, so fiddling the values of c_i is not a proper explanation for its value.

Chapter 12

Future Prospects

Most physicists would agree that the theory of quantum fields and particles has reached something of an impasse. It has become too 'comfortable', so to speak, in so far as the experimental results that pour out plentifully from the particle colliders accord pretty well with the Standard Model predictions.[1] It has been a disheartening time for budding new theorists searching for a grand overarching scheme that encompasses the known gauge theories. The reason is that none of the currently popular ideas — and some of them have been quite brilliant — such as SUSY, strings, unseen dimensions, etc., has gained any experimental support. Other grander concepts such as universes with separate landscapes, anthropic arguments, holographic perspectives are in my view more theological than physical since they cannot be checked or don't lead to new fundamental progress. Yet we are confronted by a host of puzzling facts like matter–antimatter asymmetry, the ratios of coupling constants, the nonzero but miniscule cosmological constant magnitude, the absence of magnetic monopoles, etc. They all demand sensible explanation. But none of the proposals forwarded thus far is utterly convincing or is supported by compelling observational evidence.

Given that we are in the doldrums, it is little wonder that researchers have looked for signals of new phenomena in the hope that they might shed light on the particle spectrum and the forces acting upon them. They invoke new states, such as right-handed

---

[1] Apart from a handful of weak heavy D meson decays.

gauge bosons, extra scalar fields (such as Higgs doublets or more), mirror photons, etc. Perhaps augmenting the number of fields may produce new insights that will lead to new discoveries; but at the moment they seem more like shots in the dark from a spray gun. It is proving a hard climb for aspiring new researchers because whatever they propose has not only to fit established knowledge but at the very least should lead to new vistas or alter our physical perspective dramatically.

In this monograph it has been my attitude that a return to basics is more profitable; that any progress has to depend on a complete description of events we come across and can measure, so that in addition to where and when they take place, we should be able to describe mathematically what attributes get altered. The present-day conventional approaches follow in Fermi's footsteps; they just introduce an appropriate interaction between the participants and unravel its Lorentz character by its effect on the distribution of the resulting particles. It is generally believed and accepted that the participating particles should belong to some representation of a particular Lie group and this can be used to link the behaviours of different multiplet members. Grand unified theories have exactly that objective and it must be said that they do neatly package the observed states into elegant representations. But their ambitiousness is also their downfall: by enlarging the (gauge) interactions to that of the higher group they get stuck with unpalatable processes that mix quarks with leptons, such as proton decay.

To put some mathematics into the 'what' of the event, I have invoked anticommuting complex scalar property coordinates, rather than extending spacetime by extra tiny commuting coordinates. I took inspiration from the fact that anticommuting Grassmann variables never lead to infinite modes that need to be waved/explained away. On the contrary, any states which emerge from such a scheme are always finitely constrained. I also copied Einstein in trying to geometrise the whole concept, which in turn meant mixing commuting and anticommuting coordinates as SUSY (and its N-extension) does. The way to do this was pioneered by Berezin and DeWitt as well as Arnowitt and Nath [58], but in the context of supergravity.

I have just adapted their procedures and shoe-horned them into our own notation.

Ideas come and go, but few remain entrenched into the history; those that survive punctuate the development of physics. Some ideas have had their moment of glory, or more correctly popularity, ultimately to disappear because of some weakness or disagreement with observation. I am well aware that this fate may well befall the proposals outlined in this monograph due to some or several 'unrecoverable errors' (in computer parlance) that I may have missed. If that were to happen maybe a remnant idea might survive. Time will tell, but for the rest of this section I will stick my neck out and follow through some of the consequences of the 'when–where–what' scheme.

12.1 Hunting new states

Since the whole notion of introducing additional fermionic variables is to describe all possible events and states at one stroke, let me highlight some of the principal results. The number of such variables determine what Lie groups bear consideration. I have unashamedly mimicked the promoters of grand unification by limiting the number of extra coordinates to those absolutely needed to comprise the known leptons and quarks. That has led me to consider *five* independent ζ at the very least which can engender the groups Sp(10), SU(5) and SO(10) depending on how we manage these property variables. Doubling for parity of leptons forced us to enlarge the number of variables to seven but further research may require more or fewer such coordinates, hopefully the latter.

The finite melange of properties produces certain (selfdual) states and we have shown that they include the known generations of quarks; but the blends also include more states. The most striking results/predictions of our approach are:-

- An extra generation of lepton isodoublets. There are some inklings [61] that a fourth neutrino is needed to explain certain processes but no sign yet of a fourth charged lepton.
- The third generation of quarks seems to be quite different from the first two generations. In particular the third set of quarks may be

an isotriplet with charges $(2/3, -1/3, -4/3)$. This will mess up the unitarity of the CKM mixing matrix and ought to produce a clear signal. The existence of a quark with charge $-4/3$ ($\mathcal{X}$ say) would send an even stronger signal. Perhaps the best way to investigate both predictions would be through a muon–antimuon collider because the rise in total scattering cross-section containing it will show up immediately as will the decay products of the putative $\mathcal{X}$.

- Additional 'pure' states like down quarks of type $\zeta_c(Z_c - Z_c^2/2)(1 + Z_\ell^2/2)$ and charged leptons L of type $\zeta^4 Z_\ell$ are an embarrassment, it must be admitted.

- As the gauge fields are always accompanied by their characteristic coupling constants through the frame vectors or enlarged metric and the Lagrangians for them accompany the Einstein–Hilbert gravitational term, we have to impose a uniform normalisation on them to ensure gravitational universality. We saw that this leads to a prediction (11.26) for the weak mixing angle that is quite close to experiment after taking account of all the force fields.

- Success in unifying gravity and electromagnetism using a single (electric) ζ. Changing the Klein–Kaluza fifth coordinate from commuting to anticommuting was all that was needed!

In searching for whatever nature is hiding from us, the most favourable feature of the property scheme is we only have to hunt down a finite number of them.

12.2   Seeking new principles

The curvature of property coordinates through invariant combinations like $Z = \bar{\zeta}\zeta$ and its powers works its way everywhere without endangering gauge invariance of the forces. There seems to be no principle that specifies the coefficients c of these terms even though they are limited in number. (For instance in the marriage of electrodynamics and electromagnetism there are just two independent coefficients, as we are only dealing with one electrical property.) At a semiclassical level we can regard the c as the expectation values of some uncharged scalar field ϕ. In fact if we were to promote c to a full ϕ we would no doubt be dealing with a scalar–tensor theory of gravity, though, unlike KK theory where we have to contend with modes in the fifth coordinate, there is no further dependence on the extra coordinate ζ.

In the full characterisation of SU(3)⊗SU(2)⊗U(1) geometry we came across several c_i even after assuming that gravity is chirally symmetric; there were enough c_i that we could easily fine tune the cosmological constant Λ to equal its miniscule observed value. Our scheme does not predict that value and if we were to set $\Lambda = 0$ it would just represent some sort of constraint. If we had some principle which specified the nature of the polynomial $C(Z) = \sum_n c_n Z^n$ that would be a significant advance; but we don't. One possibility would have been to allow for nonlinear transformation of properties, like $\zeta' = \zeta F(Z)$. This would lead to

$$d\zeta' = d\zeta F + \zeta[d\bar{\zeta} \cdot \zeta + \bar{\zeta} \cdot d\zeta]F'$$

plus a similar expression for $d\bar{\zeta}'$. This would then transform the flat metric distance squared ds^2 into a complicated combination of cross-terms implicating $G_{\mu\nu}, G_{\bar{\mu}\bar{\nu}}$ as well as modifying the structure of $G_{\mu\bar{\nu}}$. In other words it muddies the waters when there is more than one property. Conversely if we were to start off with such a complicated expression for ds^2 we could exploit such transformations to coax them into the 'diagonal' type $d\bar{\zeta}C(Z)d\zeta$. But that still does not tell us what the values of the c_n are. Thus there is a need for a new tenet which specifies what the property curvature polynomial $C(Z)$ must be. This tenet might arise if we give the property sector of the metric free rein and allow for all possible scalar fields, the Higgs subset being part of the components and complications notwithstanding. The c_n would then be regarded as expectation values of the property–property scalar fields.

12.3 The dark sector

Readers will have noticed that the bosonic superfield $\Phi(x, \zeta, \bar{\zeta})$ admits several scalar fields ϕ which multiply the powers of Z when expanded in property coordinates. These Z invariants mean that the corresponding ϕ are neutral in every way, except for gravitational: they are not subject to any gauge interactions with other force fields. It is tantalising to speculate that the expectation values of these

neutral fields $\langle\phi\rangle$ (in addition to those arising in the curvature polynomial) can serve to provide mass terms for quarks and leptons.
Personally I think it is unlikely that they will do the trick because
we have to cover a huge range of values, from less than a few eV
(neutrinos) up to 10^{11} eV (top quark). In my opinion it is more likely
that the tiny neutrino masses are radiatively produced, from weak
couplings to the other massive particles, by quantum loops, though
this is unproven speculation.

12.4 Quantisation

The entire discussion thus far has steered clear of quantum effects
on the entire 'when–where–what' scheme. So the next step is to see
what quantum corrections affect semiclassical effects already outlined. Primarily there is the quantisation of the force fields themselves. To ensure unitarity after gauge fixing the simplest way is to
apply BRST methods [16] though there are other ways. In BRST
one introduce ghost fields (scalar for vector QCD or QED, vector
for tensor gravity [62]) and this is most neatly done by introduc-
ing a pair $\theta, \bar{\theta}$ of auxiliary fermionic variables, much like the ζ. One
then makes the fields $A(x, \theta, \bar{\theta})$ dependent upon them and considers
translations in an extended space, which exposes the ghost fields.

I have tried to give this approach a metric interpretation but not
succeeded satisfactorily in tying all the loose ends yet. There may
even be some contradictions which have escaped my notice. Perhaps
others may succeed in exploring byways of this travelogue and make
more of the approach. It goes without saying that this whole property
scheme may be totally misguided and will collapse for one reason or
another following the dodo into extinction; if so that would not be
without precedent. I have seen intelligent theories bite the dust on
several occasions and the ideas I have outlined in this monograph may
suffer the same fate. Nevertheless I believe that my method of trying
to specify the entirety of events through some new coordinates is
not without merit. Perhaps some of the ingredient ideas may survive
if and when my proposal approaches its final judgement, buried in
good company alongside other failed proposals.

Appendix

Supplementary Mathematics

A.1 Fourier series and integrals

Any complex function of a real variable $\phi(x)$ which is unaffected by the 'translation' $x \rightarrow x + L$ can be expressed as a discrete Fourier series:

$$\phi(x) = \sum_{n=-\infty}^{\infty} \varphi_n \, \exp(2in\pi x/L), \qquad (A.1)$$

as the reader can readily verify term by term. (Note that only if ϕ is real can one say that $\varphi_n^* = \varphi_{-n}$.) Because of the orthogonality relation:

$$\int_{-L/2}^{L/2} \exp(2in\pi x/L) \exp(-2im\pi x/L) \, dx = L\delta_{mn}, \qquad (A.2)$$

where the 'Kronecker' delta $\delta_{mn} = 1$ for $n = m$, 0 otherwise, the Fourier components can be conversely calculated via the integral:

$$\varphi_n = \frac{1}{L} \int_{-L/2}^{L/2} \phi(x) \, \exp(-2in\pi x/L) \, dx. \qquad (A.3)$$

Finally we may note that the integral over the normed square of the function can be determined in a couple of ways:

$$\frac{1}{L} \int_{-L/2}^{L/2} |\phi(x)|^2 \, dx = \sum_n |\varphi_n|^2. \qquad (A.4)$$

Now imagine that we extend the length of the repeat integral: $L \to \infty$. Because the spacing between successive intervals in the exponents at $k_n \equiv 2n\pi/L$ or $dk = 2\pi/L$ becomes infinitely small, the sum over n can be regarded as a Riemann representation of an integral. In fact interpreting $\varphi(k) \equiv L\varphi_n$, we can convert the above sums into Fourier integrals and finish up with the dual representations:

$$\phi(x) = \int_{-\infty}^{\infty} \varphi(k) \, \exp(ikx) \, dk/2\pi, \quad \varphi(k) = \int_{-\infty}^{\infty} \phi(x) \, \exp(-ikx) \, dx$$

$$(A.5)$$

for well-behaved continuous functions. We will make extensive use of such formulae at appropriate places in the text.

A.2   Potentials and Green functions

A field φ with intrinsic mass m has to obey the 'mass-shell' condition $(k^2 - m^2)\varphi(k) = 0$ in momentum space. Converting this into x-space and introducing a source field $j(x)$, this implies the inhomogeneous wave equation:

$$(\square + m^2)\phi(x) = j(x). \tag{A.6}$$

If the source is a static one with magnitude unity and is placed at the origin,[1] $j(x) = \delta^3(\mathbf{r})$ and the above equation reduces to

$$\left(-\frac{\partial}{\partial \mathbf{r}} \cdot \frac{\partial}{\partial \mathbf{r}} + m^2\right) \phi(x) = \delta^3(\mathbf{r}), \tag{A.7}$$

whose Fourier transform reads $(\mathbf{k}^2 + m^2)\varphi(\mathbf{k}) = 1$. The inverse equation in configuration space is obtained by Fourier transformation via

---

[1]Point sources at the origin $x = 0$ can be represented by the Dirac delta function $\delta(x)$ which is infinitely spiked at the origin but which vanishes everywhere else.

(A.5) and reads

$$\phi(\mathbf{r}) = \int \frac{\exp(i\mathbf{k} \cdot \mathbf{r})}{(\mathbf{k}^2 + m^2)} \frac{d^3\mathbf{k}}{(2\pi)^3} = \frac{\exp(-mr)}{4\pi r}, \qquad (A.8)$$

which corresponds to a Yukawa potential. $\phi(\mathbf{r})$ fades rapidly with distance r and in the limit as $m \to 0$, it reduces to a Coulomb potential $1/4\pi r$.

In the 4D version, the Green function Δ associated with a unit source placed at x' is defined to satisfy

$$(\Box + m^2)\Delta(x - x') = -\delta^4(x - x'). \qquad (A.9)$$

Scrutiny of (A.9) shows that we can formally regard $-\Delta$ as the inverse operator $(\Box + m^2)^{-1}$, with a corresponding Fourier transform $1/(k^2 - m^2)$ called the 'propagator'. The singularity at $k^2 = m^2$ needs specification when integrating over k. For time-ordered products[2] Δ_c is called the causal or Feynman 'propagator' and is described by a contour integral in k_0-space which avoids the problem singularity at $k_0 = \pm\sqrt{\mathbf{k}^2 + m^2}$ by adding a negative infinitesimal imaginary part $-i\epsilon$ to the mass. (In fact for a resonant state or for an unstable particle the imaginary part ceases to be infinitesimal.) If we transform back into configuration space x, one can arrive at the explicit result:

$$i\Delta_c(x) = \frac{mK_1(m\rho)}{4\pi^2\rho}; \qquad \rho \equiv \sqrt{-x^2 + i\epsilon},$$

where K_1 is a Bessel function of the second kind. Like (A.8), Δ_c damps out exponentially at spacelike x and in the massless limit $i\Delta_c(x) \to 1/4\pi^2\rho^2$.

A.3 Field interactions and perturbation theory

So far we have mostly been dealing with free fields. It is now time to consider their interactions with other fields or even self-interactions due to nonlinearity. This will lead us into the subject of perturbation

---

[2] A time-ordered product of a series of functions $\phi(t)$ means placing them in the order $\phi(t_1)\phi(t_2)\ldots\phi(t_n)$ where $t_1 > t_2 > \ldots > t_n$. If the ϕ are Grassmannian further ± 1 factors need to be included.

series and Feynman diagrams. The first task is to separate the action
into various parts. We write $\mathcal{L} = \mathcal{L}_0 + \mathcal{L}_g$, where $\mathcal{L}_0$ represents the
free field action and $\mathcal{L}_g$ stands for the mutual interactions between
the various fields, connected through coupling constants g. When g
vanish the states are provided by eigenvalues of $\mathcal{L}_0$ and created by
application of operators $a^\dagger, b^\dagger$ acting on the vacuum, associated with
the free fields.

In perturbation theory the interactions are treated in powers of g
through expectation values

$$\langle \text{final} \, | T \left[\exp \left(i \int \mathcal{L}_g \, d^4x \right) \right] | \, \text{initial} \rangle \qquad (A.10)$$

between free initial and final states, with T representing a time-
ordered expansion of the exponential. Explaining how this result is
achieved would take us too far afield so I recommend reading one or
other of the standard texts on quantum field theory [12]. To illustrate
how this works, let us consider a simple example of two scalar fields:
Φ with mass m and χ with mass μ. Suppose χ is complex and Φ
is real and consider a cubic interaction between the fields (so the
coupling constant g has the dimensions of mass). Thus

$$\mathcal{L}_0(\Phi, \chi) = [(\partial^n \Phi)(\partial_n \Phi) - m^2 \Phi^2]/2 + [(\partial^n \chi^\dagger)(\partial_n \chi) - \mu^2 \chi^\dagger \chi]$$
$$(A.11)$$

$$\mathcal{L}_g(\Phi, \chi) = g \chi^\dagger \chi \Phi. \qquad (A.12)$$

The Feynman diagrams arise by expanding in powers of g; for
example suppose we examine $\chi \chi$ elastic scattering to order g^2. Since
both initial and final states involve two annihilation and creation
operators operating on the vacuum, namely

$$\langle \text{final} \, | \sim \langle 0 | a \, a, \quad | \text{initial} \, \rangle \sim a^\dagger \, a^\dagger | 0 \rangle,$$

we have to extract four powers of χ from the exponential leaving us
with a time-ordered expectation value $\langle |T[\Phi \, \Phi]|0 \rangle$; this corresponds
to a Φ propagator Δ_c acting between two cubic $g \chi^\dagger \chi \Phi$ vertices. This
can be pictured as a *Feynman diagram*, as in Figure A.1. The use
of Feynman diagrams has greatly simplified picturing and evaluating quantum effects. There even exist now some algebraic computer

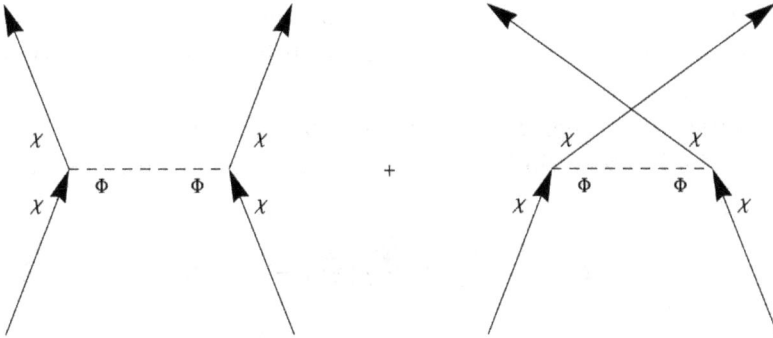

Fig. A.1 Scattering of two χ particles, mediated by Φ particle, is the sum of two Feynman diagrams.

programs that automate the process, once the starting Lagrangian density is specified. These programs are able to work out probability amplitudes for specific reactions for any Lagrangian model.

A.4 Spontaneous symmetry breaking and mass generation

We have already pointed out that the free particle Lagrangian for a complex field χ of mass μ, namely

$$\mathcal{L}_0 = \partial\chi^\dagger \cdot \partial\chi - \mu^2\chi^\dagger\chi$$

is invariant under the *global* (x-independent) phase transformations $\chi \rightarrow \exp(i\Lambda)\chi$. However as soon as we make the transformations local (x-dependent), so $\chi \rightarrow \chi' = \exp(i\Lambda(x))\chi$, the kinetic or derivative part of $\mathcal{L}_0$ acting on $\Lambda(x)$ spoils the invariance. This crucial point needs the inclusion of a gauge field to compensate the previous gauge non-invariance. If one replaces the ordinary derivative by the covariant derivative:

$$\partial\chi \rightarrow D\chi \equiv (\partial - iA(x))\chi, \quad \partial\chi^\dagger \rightarrow D\chi^\dagger \equiv (\partial + iA(x))\chi^\dagger$$

the covariance under phase transformations is restored provided the *gauge field A* picks up the slack from Λ: $A \rightarrow A' = A + \partial\Lambda$. Now when $\chi \rightarrow \chi' = \exp(i\Lambda(x))\chi$, $D\chi \rightarrow D'\chi' = \exp(i\Lambda(x))D\chi$ as well. [Essentially the same concept applies to gravity under general coordinate transformations: the gravitationally covariant derivative is invoked to account for the change of orthogonal frame from one location to another, due to metric curvature.]

The starting Lagrangian can be modified to include a quartic potential self-interaction of χ. Consider then the locally phase invariant Lagrangian

$$\mathcal{L}_g = D\chi^\dagger \cdot D\chi - g(\chi^\dagger\chi - f^2)^2/8 \qquad (A.13)$$

where g is the quartic coupling and f is a fixed mass scale parameter, producing a quadratic mass term, but with the 'wrong' sign ($m^2 = -2f^2$, as befits a tachyon), resulting from the expansion $(\chi^\dagger\chi - f^2)^2 = (\chi^\dagger\chi)^2 - 2f^2\chi^\dagger\chi + f^4$. The potential interaction in (A.13) has a minimum when $|\chi| = f$, so let us expand about that field value by writing $\chi \equiv f + \phi$, with ϕ, the quantum field, representing the disturbance about the minimum. Since $D\chi = (\partial - iA)(f + \phi) = D\phi - ifA$, the expansion of (A.13) in powers of g gives

$$\mathcal{L}_g = D\phi^\dagger \cdot D\phi + f^2 A^\mu A_\mu - g[\phi^\dagger\phi + f(\phi^\dagger + \phi)]^2/8. \qquad (A.14)$$

In conjunction with the gauge field Lagrangian of $\mathcal{L}_A = -F_{mn}F^{mn}/4$ we see that A, which started off massless, acquires a mass of f. This is basically the Higgs–Kibble [63] mechanism of spontaneously broken symmetry, once the background part of χ, i.e. $\langle 0|\chi|0\rangle = f$, becomes nonzero.

A.5 Fermions and the Dirac equation

So far I have studiously avoided treating fermions in any detail. I can no longer do so as they feature prominently in later chapters; if nothing else I will have to set out the notation. Dirac's enunciation of his equation proved a landmark in the development of quantum theory. For the first time it married special relativity and internal spin degrees of freedom in a binding contract; the equation was born out of pure thought based on notions of elegance and compactness. Dirac equation naturally predicts the existence of antiparticles and the notion of spin, without the latter appearing as an afterthought.

The modern treatment regards fermions, like electrons or protons, as the spinorial representation of the Lorentz group and the equation is nowadays written in the compact form

$$(i\gamma^n D_n - m)\psi(x) = j(x), \tag{A.15}$$

where one encounters four γ-matrices. $D_n \equiv \partial_n + ieA_n(x)$ is a covariant derivative for a fermion carrying charge $-e$, j is the fermionic current and ψ is a 4-column *spinor* like j. Equation (A.15) behaves covariantly under a local phase change $\psi(x) \rightarrow \exp(i\Lambda(x))\psi(x)$ because of the compensating effect of the electromagnetic gauge field: $eA \rightarrow eA - \partial\Lambda$.

The adjoint to ψ, defined as $\overline{\psi} = \psi^\dagger\gamma^0$ so that it transforms covariantly as well, undergoes an inverse phase change. The presence of γ^0 is simply there because of the hermiticity property: $(\gamma^n)^\dagger = \gamma^0\gamma^n\gamma^0$. Full mathematical details about gamma matrices and their properties can be accessed from many textbooks on that topic. For our purpose all we need to know is that quantisation of the fermion field can be accomplished in the same way as for Lorentz scalar fields:

$$\psi_\alpha(x) = S_{\mathbf{k}}\,[a_\alpha(\mathbf{k})\exp(-ik \cdot x) + b_\alpha^\dagger(\mathbf{k})\exp(ik \cdot x)], \tag{A.16}$$

$$\overline{\psi}^\alpha(x) = S_{\mathbf{k}}\,[a^{\dagger\alpha}(\mathbf{k})\exp(ik \cdot x) + b^\alpha(\mathbf{k})\exp(-ik \cdot x)]; \tag{A.17}$$

$$a_\alpha(\mathbf{k}) = \sum_{\lambda=\pm 1/2} a(\mathbf{k},\lambda)\,u_\alpha(k,\lambda), \quad b_\alpha^\dagger(\mathbf{k}) = \sum_{\lambda=\pm 1/2} b^\dagger(\mathbf{k},\lambda)\,v_\alpha(k,\lambda)$$

above, where u and v are solutions of helicity λ, obeying the Fourier transformed equations: $(\gamma \cdot k - m)u(k) = 0$, $(\gamma \cdot k + m)v(k) = 0$. The only noteworthy feature is that the destruction and creation operators, $a, a^\dagger$ obey *anticommutation* relations:

$$\{a(\mathbf{k},\lambda), a(\mathbf{k}',\lambda')\} = \{a^\dagger(\mathbf{k},\lambda), a^\dagger(\mathbf{k}',\lambda')\} = 0, \tag{A.18}$$

$$\{a(\mathbf{k},\lambda), a^\dagger(\mathbf{k}',\lambda')\} = \delta_{\lambda\lambda'}\overline{\delta}^3(\mathbf{k},\mathbf{k}'), \tag{A.19}$$

and similarly for the antiparticle operators $b, b^\dagger$.

When a fermion is exchanged between two locations, as in Feynman diagrams, and following the arguments in Section A.2, one comes across the 'fermion propagator',

$$\langle 0|T[\psi(x)\psi(x')]|0\rangle = i \int d^4k\, i(\gamma \cdot k - m)^{-1} \exp(ik \cdot (x - x'))$$

or $i(\gamma \cdot k + m)/(k^2 - m^2 + i\epsilon)$ as the Fourier transform. This connects the vertices of the fields which interact with ψ and corresponds to a fermionic particle carrying momentum k.

A.6 Veneziano dual resonance model

Veneziano discovered a simple function which captured the idea that a scattering amplitude, such as $p_1 + p_2 \rightarrow p_3 + p_4$ can either be expressed as a sum over resonances in one channel or in the crossed channel. See Figure A.2. In order to understand how both pictures can be combined in one formula, let us make a small mathematical detour.

The Gamma or factorial function can be represented by the convergent integral,

$$\Gamma(z) \equiv \int_0^\infty t^{z-1} \exp(-t)\, dt. \tag{A.20}$$

Integration of (A.20) by parts can be used to establish the recurrence relation, $\Gamma(z + 1) = z\Gamma(z)$. And since one knows directly from (A.2) that $\Gamma(1) = 1$, we can work up to the conclusion that for $z = n$ integer, $\Gamma(n + 1)$ is nothing but $n!$. Conversely, working downwards in z one will find that the integral for $\Gamma(z)$ becomes singular when

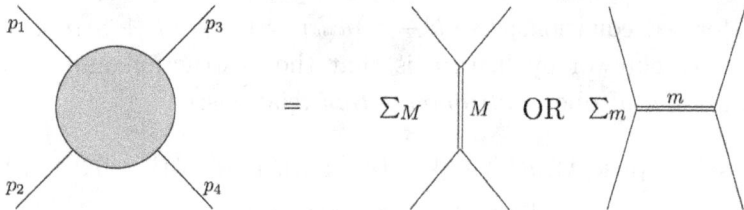

Fig. A.2 Two-particle scattering can be described as a sum of poles/resonances in the direct or crossed channels.

$z = 0, -1, -2, \ldots$. Stated differently $\Gamma(z)$ develops poles at zero and negative integer values.

This is where we can make contact with scattering amplitudes, which also contain dynamical poles in direct or crossed channels. Veneziano [64] proposed the following clever formula for the scattering amplitude M, simultaneously covering the direct and crossed poles, namely:

$$M(p_1, p_2, p_3, p_4) = \frac{\Gamma(1 - \sigma)\Gamma(1 - \tau)}{\Gamma(2 - \sigma - \tau)}, \qquad (A.21)$$

where the Lorentz invariant $\sigma = (p_1 + p_2)^2/m^2$ and $\tau = (p_1 - p_3)^2/M^2$. We can immediately understand how the amplitude will contain poles when either σ or τ takes on the values 1,2,3, etc. This crafty idea has been generalised to cover N-particle interactions with poles in various channels, but for us the important point is that it proved to be the seed that led to the development of string theory.

Glossary

Abelian groups: Groups with operation that can be carried out in any order without affecting the final result. The generators of the corresponding Lie algebras, $A, B \ldots$ thereby commute with one another: $[A, B] = AB - BA = 0$, etc. The matrices for representations of those algebras commute as well. With non-Abelian groups, such is not the case. Rather, the generators form a closed, non-commuting system.

AdS/CFT correspondence: A relation between anti-de Sitter gravitational theory and conformal or scaleless field theory, first found by Maldacena.

Betatrons: Accelerators of electrons, primarily along circular vacuum tubes using alternating current and orthogonal magnetic field.

Boson: Particle with integer spin such as a pion, gluon, photon or graviton.

Calabi–Yau manifolds: Surfaces in any number of complex dimensions which are 'Ricci-flat'. They can be quite complicated and are invoked in superstring theory where one encounters myriads of possibilities.

Chirality projections: These are either positive or negative. In the γ matrix algebra they correspond to the action of the operators $P_\pm = (1 \pm i\gamma_5)/2$ where γ_5 is the product of the four basic vector γ matrices.

CKM mixing matrix: The Cabibbo–Kobayashi–Maskawa 3×3 unitary mixing matrix which describes how the down quark mass eigenvalues are related to their flavour properties. There is a counterpart mixing for the three clearly established neutrinos.

Cosmic microwave background: The radio noise pervading outer space and it represents the cooled photon background soup from an earlier stage of cosmic evolution. It was discovered by Penzias and Wilson, almost by accident.

CP symmetry: The combination of carrying out spatial reflection or parity (P) and changing particles to antiparticles or 'charge conjugation' (C). CP was thought to be conserved in weak interactions even after the overthrow of P symmetry. Alas, CP too is violated in weak decays of certain uncharged heavy strange and beauty mesons. The violation of CP symmetry finds a natural accommodation in the mixing of down quarks and leptons of which there are at least three members.

Cyclotrons: These accelerators were first constructed at Berkeley by Lawrence. Charged particles are confined between two kidney shaped pieces and follow circular paths due to a magnetic field applied perpendicularly to their trajectories. Potential differences between the pieces are carefully alternated to accelerate their motion and timed to match the circular paths as they move between one kidney container and the other.

Equipartition theorem: It states that in an ensemble of identical systems at temperature T the average energy per member of the ensemble is $\ell T/2$ for each degree of freedom that gets excited (at large enough T). Quantum theory modifies the result at lower T because the separation of energy levels can affect the sharing of energy.

Equipotentials: 2D surfaces within 3D where the scalar potential assumes the same value.

Euler–Lagrange equations: These arise from functional integrals whose extrema we wish to find. The classic examples are the brachistochrone problem and Fermat's principle of least time. Hamilton used the principle in connection with Lagrangians.

Fermion: Particle with odd half-integer spin such as an electron, proton, neutron or neutrino.

Fine structure constants: They are represented by the symbol α and equal $g^2/4\pi$ where g is the associated coupling constant. They were originally defined for electromagnetism and determine the separation of energy levels connected with higher powers of square of the electric charge due to quantum effects.

Gauss' law: It states that the total electric/gravitational flux passing through an equipotential surface is equal to the charge/mass enclosed by the surface. The differential form of this law is that the 'divergence' of the electric/gravitational field equals the charge/mass density at any point.

General covariance: This stipulates that the equations of motion take the same general form irrespective of the observer's coordinates so as to conform to the principle of relativity.

Gluons: Quanta of chromodynamics fields. There are eight of them, amalgams of two colours.

Hadrons: Particles which interact strongly with one another, but which also enjoy weaker interactions such as electromagnetism. They can further be subclassified into integer spin 'mesons' and half integer spin 'baryons'.

Helicity: The spin along the direction of motion — forwards is positive or right helicity; backwards the helicity is left. Helicity is of course quantised, being a spin component along a particular direction.

Irreducible representation: A set of components on which the operators act as matrices and which cannot be decomposed into smaller independent parts.

Isospin: It is defined by analogy with spin and is very useful in describing the properties of nuclei. Disregarding the small mass difference and electrical properties, the proton and neutron can be considered as two isospin states of a nucleon, just like the spin up, spin down states of an electron. An $SU(2)$ isospin group acts on them and it is considered to be an almost exact invariance symmetry of nuclear interactions.

Laser interferometer: It consists of two perpendicular arms with reflecting mirrors at each end. A laser beam is split into two parts; each part travels along each arm and the beams are reflected and rejoined near the bases of each arm causing an interference pattern. The Michelson–Morley interferometer was just such an instrument but did not have the luxury of laser beams at that time.

Leptons: Particles which do *not* interact strongly and are generally light. They can be charged like electrons and muons or possess zero charge like neutrinos.

LHC: The Large Hadron Collider at CERN. Collisions between protons, nudged at various places along their huge circular paths due to intense magnetic field (produced by superconducting coils), at about $10\,\text{TeV}$ give rise to large numbers of new particles requiring special detectors and very fast electronic analysis.

Linac: Linear accelerator. The most famous of these is the SLAC linac at Menlo Park. Particles are injected into a series of linear vacuum tubes whose voltages alternate as the particles proceed and are accelerated from one tube to the next.

Liquid drop model: When many protons and neutrons cohere in a nucleus of high atomic number Z and mass A Weizsäcker proposed that they behave altogether like a liquid drop. On that basis a semi-empirical mass formula was worked out which takes account of its size, the Coulomb repulsion between protons, the surface tension effect plus smaller effects arising from the exclusion principle and the known tendency for nucleons of the same type wanting to pair up.

Magnetic monopoles: A point charge is an electric monopole. There is no counterpart in magnetism. Cabrera has searched for a magnetic monopole but without any success, apart from a single anomalous event recorded in 1982. This points to a fundamental difference between electricity and magnetism.

Muons: Heavy twins of electrons, but about 200 times as heavy and therefore unstable. Muons can be long-lived when moving near the speed of light due to relativistic time dilation. They are secondary particles produced from the decays of charged pions, themselves produced from cosmic rays. They have a further sibling in the heavier tauon of the third generation.

Neutrino transmutations: The neutrino mass states are a mixture of neutrino flavour states. The neutrinos flavour states $(\nu_e, \nu_\mu, \nu_\tau)$ can change into one another in time as they move through space. This goes into explaining why the electron neutrino flux arriving on Earth is only about one-third of the expected value, were there no transmutation.

Parity: Symmetry of space reflection. A polar or normal vector for instance will be reversed if one reflects it in a mirror: it has normal or positive parity. (That would be like an electric field.) On the other hand an axial vector like angular momentum or spin does not change sign like a polar vector would do: it has negative parity. (That would be like a magnetic field.) It is worth noting that Maxwell's equations behave uniformly under space reflection, from which one concludes that the laws of electromagnetism conserve parity.

Quantum anomalies: Anomalies in equations of motion were first discovered in the context of the divergence of the axial current. They arise from quantum loop contributions to terms that would not normally be present in the classical equations of motion. They have the potential to destroy calculability of 'renormalisable' theories, like the Standard Model.

Quantum loops: They arise in higher orders of perturbation theory. They were instrumental in establishing QED through the Lamb shift and the anomalous magnetic moment of the electron.

Quantum numbers: Eigenvalues of operators (associated with some group) in quantum theory. Examples are charge Q, baryon number B, hypercharge Y, spin S, etc. Some quantum numbers are absolutely conserved in a reaction, like Q and B; others are occasionally broken, like Y in weak interactions.

Quarks: Fundamental spin 1/2 particles with curious 1/3 integer charges. The name 'quark' derives from *Finnegan's Wake* by James Joyce and was so coined by Gell-Mann. Zweig alternatively named them 'aces' (as there were four leptons known at the time) but that term never caught on. Quarks are undetectable directly but can only be inferred indirectly from experiment. Most matter are made up of two types of quarks: up quarks carry charge $2/3$ and down quarks have charge $-1/3$. If we include the more exotic ones, there are also charm quarks, strange quarks, top quarks and bottom quarks. They come in 3 types of colour and they are enslaved in bigger hadronic combinations through the action of chromodynamic forces.

Radioactivity: There are three types. The alpha type occurs when an unstable nucleus decays into a different stabler nucleus with emission of a He nucleus or α particle; the most famous example is the decay of radium into radon. The beta minus type arises when a neutron changes into a proton, accompanied by an electron and an antineutrino (which is very hard to detect); a good example is the decay of ^{14}C into ^{14}N often used in carbon dating. The gamma type happens when, as a result of a previous decay, the daughter nucleus is formed in an excited state and subsequently decays into its nuclear ground state by emitting a high energy, very penetrating, photon or gamma particle.

Spin: The intrinsic angular momentum carried by a particle, aside from any orbital angular momentum due to the particle's orbital motion. In quantum theory, angular momentum is quantised in half-units of $\hbar$, unlike classical theory where it could be anything in principle.

Spin-statistics theorem: It stipulates that boson states do not change sign when any two identical bosons are exchanged, whereas fermion states reverse sign when identical fermions are exchanged.

Superdeterminant: If we write a matrix spanning commuting and anticommuting variables in the form $\begin{bmatrix} A & B \\ C & D \end{bmatrix}$, where the matrix A straddles the Bose–Bose sector, B the Bose–Fermi sector, C the Fermi–Bose sector and D the Fermi–Fermi sector, then the superdeterminant can be worked out as the product $\det(A - BD^{-1}C) \cdot \det(D)^{-1}$.

Supergravity; or the acronym SUGRA is a generalisation of supersymmetry (SUSY) that encompasses gravity. In such a scheme the spin 2 graviton is accompanied by a spin 3/2 'gravitino'.

Tensors: They behave in the same way as multicomponents of direct vector products. Thus under changes of coordinates $T_{klmn...}$ has the same transformation properties as $A_k B_l C_m D_n \dots$ where $A, B, C, D, \dots$ are vectors. In computer coding one can regard T as a multidimensional array.

(V–A) theory: Vectors come in two types: pure polar (V like displacement **r**) or pure axial (A like spin). The former change sign under space reflection or parity, whereas the latter do not. The combination (V–A) is what the weak interaction favours between the two currents that enter in the Fermi contact term; thus in beta decay it is a product of $j_{\text{left}}^{\text{hadron}}$ with $j_{\text{left}}^{\text{lepton}}$ because (V–A) connotes left helicity. The nature of this interaction took a good while to disentangle even though it was known that parity invariance is broken.

Vielbein: Vierbeins in 4D. They are essentially the inverses of the frame vectors and help to transcribe Lagrangians from flat space into curved space so as to ensure general covariance. In the text they can be recognised as $e^a{}_m$.

Young tableau: A pictorial way of looking at an irreducible representation which describes the permutation symmetry of its components. Thus a completely symmetrical representation looks like an aligned row of boxes, whereas an antisymmetric representation looks like a column of boxes. Mixed symmetry representations consist of a column of rows with diminishing size.

Zeeman effect: The splitting of spectral lines that occur when the sources of light are subject to strong magnetic fields. They point to the existence of slightly separated energy levels, associated with angular momentum. The effect was first seen in the spectral lines produced by the Sun and played a role in the discovery of helium.

Bibliography

[1] G. Zweig (2010), arXiv:1007.0494; M. Gell-Mann, Phys. Lett. **8**, 214 (1964).

[2] A. Salam, Nuovo Cimento **5**, 299 (1957).

[3] T.D. Lee and C.N. Yang, Phys. Rev. **104**, 254 (1957).

[4] C.S. Wu, Phys. Rev. **105**, 1413 (1957).

[5] E.C.G. Sudarshan and R.E. Marshak, Phys. Rev. **109**, 1860 (1958).

[6] R.P. Feynman and M. Gell-Mann, Phys. Rev. **109**, 193 (1958).

[7] M. Longair, Proc. Amer. Phil. Soc. **155**, 147 (2011).

[8] R.A. Hulse and J.H. Taylor, Astrophys. J. **195**, L51 (1975).

[9] LIGO and Virgo Collaboration, Phys. Rev. Lett. **116**, 061102 (2016).

[10] D.J. Gross and F. Wilczek, Phys. Rev. Lett. **30**, 1343 (1973); H.D. Politzer, Phys. Rev. Lett. **30**, 1346 (1973).

[11] M. Fierz, Helv. Physica Acta. **12**, 3 (1939); W. Pauli, Phys. Rev. **58**, 716 (1940).

[12] S. Weinberg, *The Quantum Theory of Fields*, Cambridge University Press, Volumes 1, 2 and 3 (2019).

[13] S. Weinberg, *Gravitation and Cosmology: Principles and Applications of the General Theory of Relativity*, Wiley (1972).

[14] S.M. Carroll, *Spacetime and Geometry: An Introduction to General Relativity*, Addison-Wesley (2004).

[15] M. Fierz and W. Pauli, Proc. R. Soc. Lond. **A 173**, 211 (1939).

[16] L.D Faddeev and V. Popov, Phys. Lett. **25B**, 29 (1967); C. Becchi, A. Rouet, and R. Stora, Ann. Phys. **98**, 287 (1976); I.V. Tyutin, Lebedev Physics Institute preprint 39 (1975), arXiv:0812.0580.

[17] B. DeWitt, *Supermanifolds*, Cambridge University Press (1992).

[18] F.A. Berezin, *The Method of Second Quantization*, Academic Press NY (1966).

[19] A. Rogers, *Supermanifolds: Theory and Applications*, World Scientific (2007).

[20] I.G. Halliday and R.M. Ricotta, Phys. Lett. **193B**, 241 (1987); G.V. Dunne and I.G. Halliday, Phys. Lett. **193B**, 247 (1987).

[21] R.C. King, Can. J. Math. **33**, 176 (1972).

[22] L.M. Krauss, *A Universe from Nothing: Why There is Something Rather than Nothing*, Free Press (2012).

[23] C.N. Yang, R.L. Mills, Phys. Rev **96**, 191 (1954).

[24] M. Gell-Mann, in *The Eightfold Way: A Theory of Strong Interaction Symmetry*, ed. M. Gell-Mann and Y. Neeman, W.A. Benjamin (1964).

[25] Th. Kaluza. Sitz. ber. Preuss. Akad. Wiss. Berlin. (Math. Phys.) **1921**, 966 (1921); O. Klein, Zeit. fur Physik **A37**, 895 (1921).

[26] T. Appelquist, A. Chodos, P.G.O. Freund, *Modern Kaluza–Klein Theories*, Addison-Wesley, Menlo Park (1987).

[27] S.L Glashow, Nucl. Phys. **22**, 579 (1961); A. Salam and J.C. Ward, Phys. Lett. **13**, 168 (1964); S. Weinberg, Phys. Rev. Lett. **19**, 1264 (1967); A. Salam, in *Eighth Nobel Symposium*, ed. N. Svartholm, Almquist and Wiksell (1968).

[28] H. Georgi and S. Glashow, Phys. Rev. Lett. **32**, 438 (1974).

[29] J.C. Pati and A. Salam, Phys. Rev. **D10**, 275 (1974).

[30] H. Fritzsch and P. Minkowski, Ann. Phys. **93**, 193 (1975).

[31] D.V. Volkov and V.P. Akulov, Phys. Lett. **46B**, 109 (1973).

[32] J. Wess and B. Zumino, Nucl. Phys. **70B**, 39 (1974).

[33] A. Salam and J. Strathdee, Fortsch. Phys. **26**, 57 (1978).

[34] L. O'Raifeartaigh, Phys. Rev. **139B**, 1052 (1965).

[35] S. Coleman and J. Mandula, Phys. Rev. **159**, 1251 (1967).

[36] S. Deser and B. Zumino, Phys. Lett. **B62**, 335 (1976); D.Z. Freedman, P. van Nieuwenhuizen and S. Ferrara, Phys. Rev. **D13**, 3214 (1976).

[37] Y. Nambu, *Lectures on the Copenhagen Summer Symposium* (1970).

[38] A. Polyakov, Phys. Lett. **B103**, 207 (1981).

[39] S. Deser and B. Zumino, Phys. Lett. **65**, 369 (1976).

[40] P. Ramond, Phys. Rev. **D3**, 2415 (1971).

[41] A. Neveu and J.H. Schwarz, Phys. Lett. **34B**, 517 (1971).

[42] M.B. Green and J.H. Schwarz, Phys. Lett. **B149**, 117 (1984).

[43] M.B. Green, J.H. Schwarz and E. Witten, *Superstring Theory*, Cambridge University Press (1987).

[44] J. Polchinski, *Strings*, Vol. I and II, Cambridge University Press (2011).

[45] R.D. Peccei and H.R. Quinn, Phys. Rev. Lett. **38**, 1440 (1977); Phys. Rev. **16D**, 1791 (1977).

[46] C. Rovell, *Quantum Gravity*, Cambridge University Press (2004).

[47] L. Smolin, *The Structural Foundations of Quantum Gravity*, Clarendon Press (2006).

[48] B. Carter and W.H. McCrea, Phil. Trans. Royal Soc. **A310**, 347 (1983).

[49] J.D. Barrow and F.J. Tipler, *The Anthropic Cosmological Principle*, Oxford University Press (1988).

[50] S. Weinberg, Phys. Rev. Lett. **59**, 2607 (1987).

[51] S. Hossenfelder, *Lost in Math: How Beauty Leads Physics Astray*, Basic Books (2018).

[52] C.B. Thorn, *Sakharov Conference on Physics*, Moscow, p. 447 (1994).

[53] R. Bousso, Rev. Mod. Phys. **74**, 825 (2002).

[54] L. Susskind, J. Math. Phys. **36**, 6377 (1995).

[55] J. Maldacena, Adv. Theor. Math. Phys. **2**, 231 (1998).

[56] A. Barducci, F. Buccella, R. Casalbuoni, L. Lusanna and E. Sorace, Phys. Lett. **67B**, 344 (1971); R. Casalbuoni and R. Gatto, Phys. Lett. **88B**, 306 (1979).

[57] R. Delbourgo, Int. J. Mod. Phys. **A31** 1650153 (2016); P.D. Stack and R. Delbourgo, Int. J. Mod. Phys. **A30**, 1550005 (2015).

[58] R.P. Nath and R. Arnowitt, Phys. Lett. **B56**, 177 (1975).

[59] M. Asorey and P.M. Lavrov, J. Math. Phys. **50**, 013530 (2009).

[60] C.H. Brans and R.H. Dicke, Phys. Rev **124**, 925 (1961).

[61] A.A. Aguilar-Arevalo *et al.* (MiniBooNE Collaboration), Phys. Rev. Lett. **121**, 221801 (2018).

[62] R. Delbourgo and M. Ramon-Medrano, Nucl. Phys. **B110**, 467 (1976).

[63] F. Englert and R. Brout, Phys. Rev. Lett. **13**, 321 (1964); P.W. Higgs, Phys. Rev. Lett. **13**, 508 (1964); G.S. Guralnik, C.R. Hagen and T.W.B. Kibble, Phys. Rev. Lett. **13**, 585 (1964); T.W.B. Kibble, Phys. Rev. **155**, 1554 (1967).

[64] G. Veneziano, Nuovo Cim. **A57**, 190 (1968); L. Susskind, Phys. Rev **D1**, 1182 (1969); Z. Koba and H.B. Nielsen, Nucl. Phys. **B10**, 633 (1969).

www.ingramcontent.com/pod-product-compliance
Lightning Source LLC
Chambersburg PA
CBHW050629190326
41458CB00008B/2200